AF655368

Basics of
INTEGRAL CALCULUS

Basics of
INTEGRAL CALCULUS

Rohit Garg

ANMOL PUBLICATIONS PVT. LTD.
NEW DELHI-110 002 (INDIA)

ANMOL PUBLICATIONS PVT. LTD.
Regd. Office: 4360/4, Ansari Road, Daryaganj,
New Delhi-110002 (India)
Tel.: 23278000, 23261597, 23286875, 23255577
Fax: 91-11-23280289
Email: anmolpub@gmail.com
Visit us at: www.anmolpublications.com

Branch Office: No. 1015, Ist Main Road, BSK IIIrd Stage
IIIrd Phase, IIIrd Block, Bengaluru-560 085 (India)
Tel.: 080-41723429 • Fax: 080-26723604
Email: anmolpublicationsbangalore@gmail.com

Basics of Integral Calculus

© Reserved

First Edition, 2011

ISBN 978-81-261-4837-0

PRINTED IN INDIA

Printed at Suman Printers, Delhi

Contents

Preface

The major advance in integration came in the 17th century with the independent discovery of the fundamental theorem of calculus by Newton and Leibniz. The theorem demonstrates a connection between integration and differentiation. This connection, combined with the comparative ease of differentiation, can be exploited to calculate integrals. In particular, the fundamental theorem of calculus allows one to solve a much broader class of problems. Equal in importance is the comprehensive mathematical framework that both Newton and Leibniz developed. Given the name infinitesimal calculus, it allowed for precise analysis of functions within continuous domains. This framework eventually became modern calculus, whose notation for integrals is drawn directly from the work of Leibniz.

Integrals appear in many practical situations. Consider a swimming pool. If it is rectangular with a flat bottom, then from its length, width, and depth we can easily determine the volume of water it can contain, the area of its surface, and the length of its edge. But if it is oval with a rounded bottom, all of these quantities call for integrals.

Author

1

Integral Calculus

Integration is the reverse of differentiation. With differentiation breaking a process down to look at the instantaneous process, integration sums up the information from small time intervals to give a total result over a larger time period.

One example of this is the area under the plasma concentration time curve. Later we shall learn that this summation or integration process can be used to evaluate dosage forms, that is it can be used as a measure of performance. Using integral calculus we can go in the reverse direction. Among the limiting processes of analysis, there are two processes with an especially important role, not only because they arise in many different connections, but chiefly due to their very close reciprocal relationship. Isolated examples of these two limiting processes, differentiation and integration, have even been considered in classical times; but it is the recognition of their complementary nature and the resulting development of a new and methodical mathematical procedure which marks the beginning of the real systematic differential and integral calculus.

The credit of initiating this development belongs equally to the two great geniuses of the Seventeenth Century, Newton and Leibnitz, who, as we know to-day, made their discoveries independently of each other. While Newton, in his investigations, may have succeeded in stating his concepts more clearly, Leibnitz's notation and methods of calculation were more highly developed; even to-day, these formal portions of Leibnitz's work form an indispensable element in the theory. We first encounter the integral in the problem of measuring the area of a plane region bounded by curved lines. Then, more refined considerations permit us to separate the notion of integral from the intuitive idea of area and to express it analytically in terms of the notion of number only.

We shall find this analytical definition of the integral to be of great significance not only because it alone enables us to attain complete clarity in our concepts, but also because its applications extend far beyond the calculation of areas. We shall begin by considering the matter intuitively.

THE ANALYTICAL DEFINITION OF THE INTEGRAL

In the above section, we have considered the definite integral as a number given by an area, hence to a certain extent as previously known, and have subsequently represented it as a limiting value. We shall now reverse this procedure. We no longer take the point of view that we know by intuition how an area can be assigned to the region under a continuous curve or, indeed, that this is possible; we shall, on the contrary, begin with sums formed in a purely analytical way, like the upper and lower sums defined previously, and shall then prove that these sums tend to a definite limit.

We take this limiting value as the definition of the integral and of the area. We are naturally led to adopt the formal symbols which have been used in the integral calculus since Leibnitz's time. Let $f(x)$ be a function which is positive and continuous in the interval $a \leq x \leq b$ (of length $b - a$). We think of the interval as being sub-divided by (n—1) points $x_1, x_2, \ldots\ldots, x_{n-1}$ into n equal or unequal sub-intervals and, in addition, we let $x_0=a$, $x_n=b$. In each interval, we choose a perfectly arbitrary point, which may be within the interval or at either end; suppose that in the first interval we choose the point x_1, in the second one the point x_2, and in the n-th interval the point x_n.

Instead of the continuous function $f(x)$, we now consider a discontinuous function (step-function) which has the constant value $f(\xi_1)$ in the first sub-interval, the constant value $f(\xi_2)$ in the second sub-interval, the constant value $f(\xi_n)$ in the n-th sub-interval. As is shown in Fig., the graph of

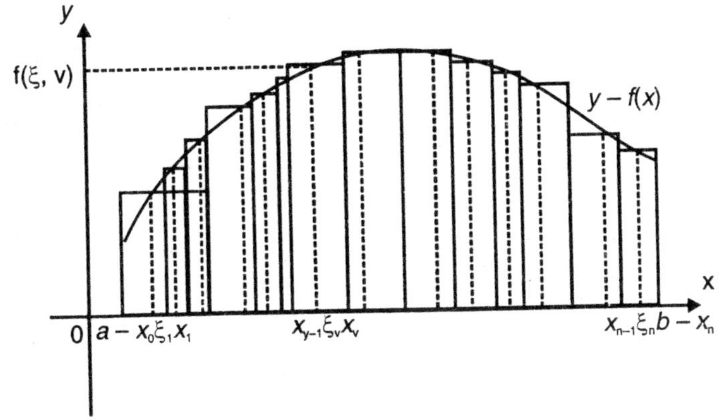

Fig. To Illustrate the Analytical Definition of Integral

this step-function defines a series of rectangles, the sum of the areas of which is given by

$$f_n = (x_1 - x_0) f(\xi_1) + (x_2 - x_1) f(\xi_2) + \dots + (x_n - x_{n-1}) f(\xi_n)$$

This expression is usually abbreviated by means of the summation sign Σ:

$$f_n = \sum_{y=1}^{n} (x_v - x_{v-1}) f(\xi_v);$$

by introducing the symbol

$$\Delta x_v = x_v - x_{v-1}$$

we can simplify this formula to

$$F_n = \sum_{v=1}^{n} f(\xi_v) \Delta x_v$$

(Here the symbol Δ is not a factor, but denotes the word difference. By definition, the entire inseparable symbol Δx_v means the length of the ν-th sub-interval.) Our basic assertion may now be stated as follows: If we let the number of points of sub-division increase without limit and at the same time let the length of the longest sub-interval tend to 0, then the above sum tends to a limit. This limit is independent of the particular manner in which the points of division $x_1, x_2, \ldots, x_{n-1}$ and the intermediate point $\xi_1, \xi_2, \ldots, \xi_{n-1}$ are. chosen.

We shall call this limiting value the definite integral of the function $f(x)$, the integrand, between the limits a and b; as we have already mentioned, we shall consider it as the definition* of the area under the curve $y = f(x)$ for $a\ x\ b$. Our basic assertion may then be re-worded: If $f(x)$ is continuous in $a \leq x \leq b$, its definite integral between the limits a and b exists.

Of course, we may also define the notion of are in a purely geometrical way and then prove that such a definition is equivalent to the above limit=definition. This theorem on the existence of the definite integral of a continuous function can be proved by purely analytical methods and without appeal to intuition.

We shall nevertheless pass it over for the present and return to it in after the use of the concept of integral has stimulated the reader's interest in constructing for it a firm foundation. For the moment, we shall content ourselves with the fact that the intuitive considerations above have made the theorem appear to be extremely plausible.

THE INTEGRAL AS AN AREA

Let there be are given a function $f(x)$, which is continuous and positive in an interval, and two values a and b $(a < b)$ in that interval. We think of the function as being represented by a curve and consider the area of the region which is bounded above by the curve, at the sides by the straight lines $x = a$ and $x = b$ and below by the portion of the x-axis between the points a and b.

That there is a definite meaning to speaking of the area of this region is an assumption inspired by intuition, which we state here expressly as a hypothesis.

We call this area F_a^b the definite integral of the function $f(x)$ between the limits a and b. When we actually seek to assign a numerical value to this area, we find that we are, in general, unable to measure areas with curved boundaries, but we can measure polygons with straight sides by dividing them into rectangles and triangles. Such a sub-division of our area is usually impossible. It is, however, only a short step to conceive in the following manner the area as the limiting value of a sum of areas of rectangles.

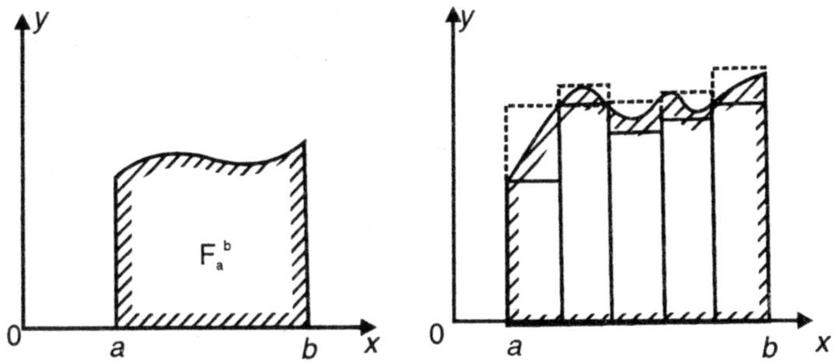

Fig. Upper Sum and Lower Sum

We subdivide the part of the *x*-axis between *a* and *b* into *n* equal parts and erect at each point of sub-division the ordinate up to the curve; the area is thus divided into *n* strips.

We can no more calculate the area of such strips than we could that of the original surface; but if, we find first the least and then the greatest value of the function $f(x)$ in each sub-interval and then replace the corresponding strip by a rectangle with height equal to the least value of the function, and by a rectangle with height equal to the greatest value of the function, we obtain two step-shaped figures. The first step-shaped figure obviously has an area which is at most equal to the area F_a^b which we are trying to determine; the second has an area which is at least as large as F_a^b. If we denote the sum of the areas of the first set of rectangles by $\underline{F_n}$ (lower sum) and the sum of the areas of the second set by $\overline{F_n}$ (upper sum), we find

$$\underline{F_n} \leq F_a^b \leq \overline{F}_n$$

If we now make the subdivision finer and finer, i.e., let *n* increase without limit, intuition tells us that the quantities $\overline{F_n}$ and $\underline{F_n}$ approach closer and closer to each other and tend to the same limit F_a^b. We may therefore consider our integral as the limiting value

$$F_a^b = \lim_{x\to\infty} \underline{F_n} = \lim_{x\to\infty} \overline{F_n} .$$

Intuition also tells us the possibility of an immediate generalization. It is by no means necessary that the n sub-intervals should all be of the same length. On the contrary, they may have different lengths provided only that, as n increases, the length of the longest sub-interval tends to 0.

FUNDAMENTAL RULES

The above definition of the integral as the limit of a sum led Leibnitz to express the integral by the symbol

$$\int_a^b f(x)dx$$

The integral symbol is a modification of a summation sign which has the shape of a long S. The passage to the limit from a sub-division of the interval into finite portions Δx_v is suggested by the use of the letter d in place of Δ.

However, we must guard ourselves against thinking of dx as an infinitely small quantity or infinitesimal, or of the integral the sum of an infinite number of infinitely small quantities. In the above figures, we have assumed: that the function $f(x)$ is positive throughout the interval, and that $b > a$.

However, the formula which defines the integral as the limit of a sum is independent of any such assumptions. For if $f(x)$ is negative in all or a part of our interval, the only effect is to make the corresponding factors $f(\xi_v)$ in our sum negative instead of positive.

We shall naturally assign to the region bounded by the part of the curve below the x-axis a negative area, which is in agreement with the familiar convention of sign in analytical geometry. The total area bounded by a curve will thus, in general, be the sum of positive and negative terms, corresponding to the portions of the curve above and below the x-axis, receptively.

If we also omit the condition $a < b$ and assume that $a > b$, we can still retain our arithmetic definition of integral; the only change is that, if we traverse the interval from a to b, the differences Δx_v are negative. We are thus led to the relation

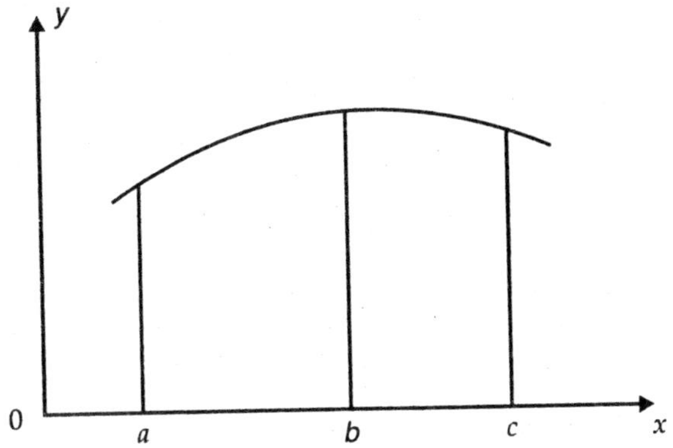

$$\int_a^b f(x)\,dx = -\int_a^b f(x)\,dx$$

which holds for all values of a and b ($a > b$). Correspondingly, we define $\int_a^b f(x)\,dx$ as equal to zero. Our definition immediately yields the basic relation:

$$\int_a^b f(x)\,dx + \int_b^c f(x)\,dx = \int_a^c f(x)\,dx$$

for $a < b < c$. By means of the preceding relations, we find at once that this equation is also true for any relative positions of the end-points a, b, x. We obtain a simple but important fundamental rule by considering the function $cf(x)$, where c is a constant. From the definition of the integral, we immediately obtain

$$\int_a^b cf(x)\,dx = c\int_a^b f(x)\,dx$$

Moreover, we assert the addition rule: If

$$f(x) = \phi(x) + \psi(x)$$

then

$$\int_a^b f(x)\,dx = \int_a^b \phi(x)\,dx + \int_a^b \psi(x)\,dx$$

The proof is quite simple.

We add a final remark about the variable of integration, which is perfectly obvious, but very important in applications. We have written our integral in the form $\int_a^b f(x)\,dx$. During evaluation of the integral, it does not matter whether we use the letter x or any other letter to denote the abscissae of the co-ordinate system, i.e., the independent variable. We use this particular symbol for the variable of integration and it is therefore a matter of complete indifference; instead of $\int_a^b f(x)dx$, we could equally well write $\int_a^b f(t)dt$ or $\int_a^b f(u)du$ or any similar expression.

Examples: We are now in a position to carry out the limiting process prescribed by our definition of the integral and thus actually calculate the area in question in a number of special cases; we shall do this in a series of examples, where we shall employ only the upper or the lower sum.

INTEGRATION OF *x*

A problem which is not quite as simple as that of the integration of the function $f(x) = x$, or, in geometrical language, of determining the area of the region bounded by a segment of a parabola, a segment of the x-axis and two ordinates.

For example, consider the integral $\int_0^b x^2 dx$ where b ý 0 and sub-divide the interval $0 \leq x \leq b$ into n equal parts of length $h = b/n$; the area which we then wish to find is the limit of the expression (upper sum):

$$h\left(h^2 + 2^2h^2 + 3^2h^2 + \ldots + n^2h^2\right) = h^3\left(1^2 + 2^2 + \ldots + n^2\right)$$

$$= b^3\left(1^2 + 2^2 + \ldots + n^2\right)/n^3$$

However, the sum in brackets has already been found.

$$1^2 + 2^2 + \ldots + n^2 = \frac{1}{6}n(n+1)(2n+1)$$

If we substitute this expression and rewrite the result in a slightly different form, our sum becomes

$$\frac{b^3}{6}\left(1+\frac{1}{n}\right)\left(2+\frac{1}{n}\right)$$

As n increases beyond all bounds, this expression tends to the limit $b/3$ and we obtain the required integral

$$\int_0^b x^2 dx = \frac{1}{3}b^3$$

Using the general result above, we immediately derive the formula

$$\int_0^b x^2 dx = \int_0^b x^2 dx - \int_0^a x^2 dx = \frac{1}{3}\left(b^2 - a^3\right)$$

Integration of x^α, where α is any positive integer: As a third example, we consider the integration of the function

$$y = f(x) = x^n$$

where α is any positive integer. For the computation of the integral

$$\int_a^b x^n dx,$$

(where we assume $0 < \alpha < b$), it would be inconvenient to subdivide the interval into n equal parts.* However, the passage to the limit may be accomplished very easily, if we subdivide in geometric progression in the following manner. Let $\sqrt[n]{b/a} = q$ and subdivide the interval by the points

$$a, aq, aq^2, \ldots, aq^{n-1}, aq^n = b$$

We should then be obliged to base the evaluation of the integral upon the calculation of the limit of

$$\frac{1}{n^{\alpha+1}}\left(1^\alpha + 2^\alpha + \ldots + n^\alpha\right)$$

as n } }; the reader may work this result out as has been indicated in the footnote referred to above.

The required integral is then the limit of the sum:

$$a^\alpha (aq - a) + (aq)^\alpha \left(aq^2 - aq\right) + \left(aq^2\right)^\alpha \left(aq^3 - aq^2\right) + \ldots$$

$$+ \left(aq^{n-1}\right)^\alpha \left(aq^n - aq^{n-1}\right)$$

$$= a^{\alpha+1}(q-1)\left\{1 + q^{\alpha+1} + q^{2(\alpha+1)} + q^{3(\alpha+1)} + \ldots + q^{(n-1)(\alpha+1)}\right\}$$

The terms in the last bracket form a geometric progression with the common ratio $q^{\alpha+1}$ } 1. If we sum this progression, we obtain for the entire expression the value

$$a^{\alpha+1}(q-1)\frac{q^{n(\alpha+1)}-1}{q^{\alpha+1}-1}$$

We now replace q by its value $(b/a)^{1/n}$; our sum then takes the form

$$\left(b^{\alpha+1}-a^{\alpha+1}\right)\frac{q-1}{q^{\alpha+1}-1}$$

If we now let n increase without limit, the first factor retains its value. Since q } 1, we can use the formula for the sum of a geometric progression and write the second factor in the form

$$\frac{1}{q^{\alpha}+q^{\alpha-1}+\ldots+1};$$

as the equation $q = (b/a)^{1/n}$ shows that q tends to 1 as n } }, the second factor will have the limit $1/(\alpha + 1)$. Thus, finally, the value of our integral is given by

$$\int_a^b x^a dx = \frac{1}{\alpha+1}\left(b^{\alpha+1}-a^{\alpha+1}\right)$$

In principle,the above calculation is simple, but itsdetails are somewhat complicated. We shall later on discover that it can be entirely avoided once we have become better acquainted with integration theory.

INTEGRATION OF A LINEAR FUNCTION

We first consider the function $f(x) = x^n$, where n is an integer greater than or equal to 0. For n = 0, i.e. for $f(x) = 1$, the result is so obvious that we simply write it down:

$$\int_a^b 1dx = \int_a^b dx = b - a$$

For the function $f(x) = x$, the integration is again a triviality from the geometrical point of view. Its integral

$$\int_a^b xdx$$

is simply the area of the trapezoid which by an elementary formula is

$$\frac{1}{2}(b-a)(b+a) = \frac{1}{2}\left(b^2-a^2\right)$$

We shall now verify that our limiting process leads to exactly the same result. In calculating the limit, we can restrict ourselves to the discussion of upper sums or lower sums. We subdivide the interval from a to b into n equal parts by means of the points of sub-division

$$a+h, a+2h, ..., a+(n-1)h,$$

where $h = (b - a)/n$. The integral must then be the limit of the following sum, which is an upper sum, if $b < a$ and a lower sum if $b > a$:

$$h\{a+(a+h)+(a+2h)+...+(a+(n-1)h)\}$$

$$=h\{na+h+2h+...+(n-1)h\}$$

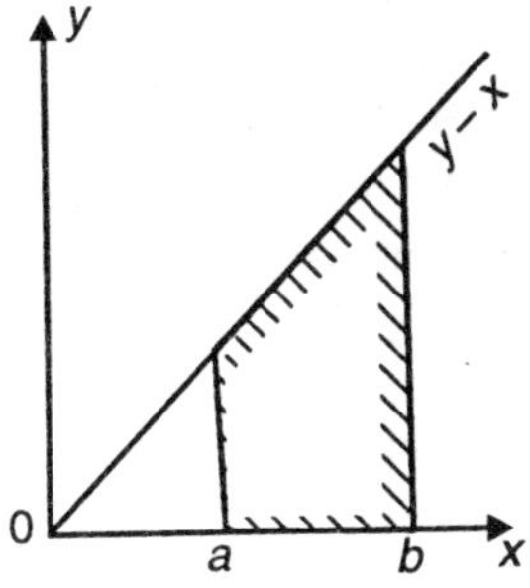

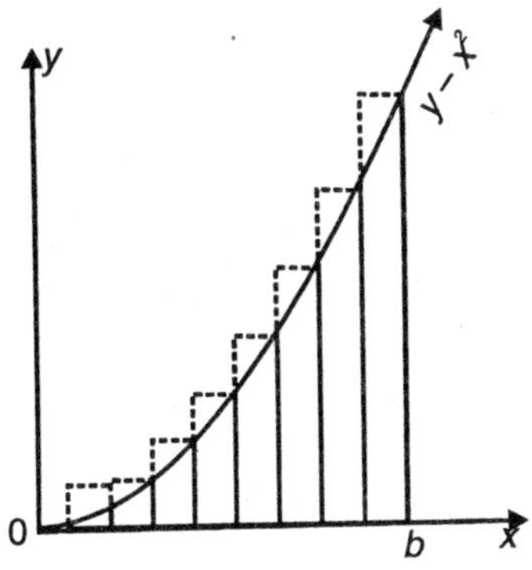

By an elementary formula, we have

$$1+2+...+(n-1)=\frac{1}{2}n(n-1)$$

and our expression may therefore be rewritten in the form

$$nh\left(a+h\frac{n-1}{2}\right) = (b-a)\left(a+\frac{b-a}{2}\frac{n-1}{n}\right)$$

As n increases, the right hand side obviously tends to the limit

$$(b-a)\left\{a+\frac{1}{2}(b-a)\right\} = \frac{1}{2}\left(b^2 - a^2\right)$$

as was to be proved.

INTEGRATION OF x^a

The result obtained above may be generalized considerably without essential complication of the method.

Let $\alpha = r/s$ be a positive rational number, r and s positive integers; in the evaluation of the integral given above, nothing is changed except the evaluation of the limit $(q-1)/(q^{n+1}-1)$ as q approaches 1.

This expression is now simply $(q-1)/(q^{(r+s)/s}-1)$. Let us set $q^{1/s} = \tau(\tau \} 1)$; then, as q tends to 1, τ will also tend to 1. We have therefore to find the limiting value of $(\tau^s - 1)/(\tau^{r+s} - 1)$ as τ approaches 1. If we divide both the numerator and denominator by $\tau - 1$ and transform them as before, the limit simply becomes

$$\lim_{\tau \to I} \frac{\tau^{3-1} + \tau^{3-2} + \ldots + 1}{\tau^{\tau+3-1} + \tau^{\tau+3-2} + \ldots + 1}$$

Since both the numerator and denominator are continuous in τ, this limit is at once determined it we set $\tau = 1$. We thus arrive at the limit $s/(r+s) = 1/(\alpha+1)$ and obtain for every positive rational value of α

$$\int_a^b x^a dx = \frac{1}{\alpha+1}\left(b^{\alpha+1} - a^{\alpha+1}\right)$$

This formula remains valid for negative rational values of α, provided we exclude the value $\alpha = -1$ for which the formula used above for the sum of the geometric progression loses its meaning. We must now investigate the limit of the expression $(q-1)/(q^{n+1}-1)$ for negative values of α, say $\alpha = -r/s$. For this purpose, we set $q^{-1/s} = \tau$ and obtain

$$q = \tau^{-s}, q^{\alpha+1} = q^{-(r-s)/s} = \tau^{r-s}$$

Hence we seek to determine the limiting value of

$$\frac{\tau^{-s}-1}{\tau^{r-s}-1}=\frac{1-\tau^{s}}{\tau^{r}-\tau^{s}}$$

We leave it to the reader to prove that this limit is again equal to $1/(\alpha+1)$, that is, that we have the integral

$$\int_a^b x^a dx=\frac{1}{\alpha+1}\left(b^{\alpha+1}-a^{a+1}\right).$$

for the general case of rational values of either positive or negative α, with the exception of $\alpha=-1$. The form of the right-hand side of this equation shows that the expression is not valid for $\alpha=-1$, since both the numerator and denominator would then be zero. It is natural to assume that the range of validity of cur last formula extends also to irrational values of α. We shall actually establish this in equation by a simple passage to the limit.

Integration of sin x and cos x: As a last example, consider the function $f(x)$ = sin x. Also in this case, we shall employ a special device. We express the integral

$$\int_a^b \sin x dx$$

as the limit of the sum:

$$S_n=h\{\sin(\alpha+h)+\sin(\alpha+2h)+\ldots+\sin(\alpha+nh)\}$$

where $h=(b-a)/n$. We multiply the right-hand bracket by 2 sin $h/2$ and recall the well-known trigonometrical formula

$$2\sin u\sin v=\cos(u-v)-\cos(u+v);$$

provided h is not a multiple of 2π, we obtain

$$\frac{h}{2\sin\frac{h}{2}}\left\{\begin{array}{c}\cos\left(\alpha+\frac{h}{2}\right)-\cos\left(\alpha+\frac{3}{2}h\right)+\cos\left(\alpha+\frac{3}{2}h\right)-\cos\left(\alpha+\frac{5}{2}h\right)\\ +\ldots+\cos\left(\alpha+\frac{2n-1}{2}h\right)-\cos\left(\alpha+\frac{2n+1}{2}h\right)\end{array}\right\}$$

$$=\frac{h}{2\sin\frac{h}{2}}\left\{\cos\left(\alpha+\frac{h}{2}\right)-\cos\left(\alpha+\frac{2n+1}{2}h\right)\right\}$$

Since $a + nh = b$, the integral becomes the limit of

$$\frac{h}{2\sin\frac{h}{2}}\left\{\cos\left(\alpha+\frac{h}{2}\right)-\cos\left(b+\frac{h}{2}\right)\right\} \text{as } h \to 0$$

Now we know from that as h tends to 0, the expression $h/2/\sin h/2$ approaches the limit 1. The desired limit is then simply $\cos a - \cos b$ and we thus arrive at the integral

$$\int_a^b \sin x dx = -(\cos b - \cos a)$$

In the same manner, the reader may verify the formula

$$\int_a^b \cos x dx = \sin b - \sin \alpha$$

Almost every one of these examples has been attacked by means of some special method or particular device. The essential point of the systematic Integral and Differential Calculus is the very fact that we utilize instead of such special devices considerations of a general character which lead us directly to the desired result. In order to arrive at these methods, we must now turn our attention to the other fundamental concept of higher analysis, the derivative.

2

Methods of Integration

Most integrals involving "simple" substitutions will not have any of the substitution work shown. It is going to be assumed that you can verify the substitution portion of the integration yourself. Also, most of the integrals done in this chapter will be indefinite integrals. It is also assumed that once you can do the indefinite integrals you can also do the definite integrals and so to conserve space we concentrate mostly on indefinite integrals. There is one exception to this and that is the Trig Substitution section and in this case there are some subtleties involved with definite integrals that we're going to have to watch out for. Outside of that however, most sections will have at most one definite integral example and some sections will not have any definite integral examples.

INTEGRATION BY PARTS

Let's start off with this section with a couple of integrals that we should already be able to do to get us started. First let's take a look at the following.

$$\int e^x dx = e^x + c$$

So, that was simple enough. Now, let's take a look at,

$$\int xe^x dx$$

To do this integral we'll use the following substitution.

$$u = x^2 \qquad du = 2xdx \qquad \Rightarrow \qquad xdx = \frac{1}{2}du$$

$$\int xe^{x^2} dx = \frac{1}{2}\int e^u du = \frac{1}{2}e^u + c = \frac{1}{2}e^{x^2} + c$$

Again, simple enough to do provided you remember how to do substitutions. By the way make sure that you can do these kinds of substitutions quickly and easily. From this point on we are going to be doing these kinds of substitutions in our head. If you have to stop and write these out with every problem you will find that it will take to significantly longer to do these problems.

Now, let's look at the integral that we really want to do.

$$\int xe^{x^2}\,dx$$

If we just had an e^{6x} by itself or by itself we could do the integral easily enough. But, we don't have them by themselves, they are instead multiplied together. There is no substitution that we can use on this integral that will allow us to do the integral. So, at this point we don't have the knowledge to do this integral. To do this integral we will need to use integration by parts so let's derive the integration by parts formula. We'll start with the product rule.

$$(fg)' = f'g + fg'$$

Now, integrate both sides of this.

$$\int (fg)'\,dx = \int f'g + fg'\,dx$$

The left side is easy enough to integrate and we'll split up the right side of the integral.

$$fg = \int f'g\,dx + \int fg'\,dx$$

Note that technically we should have had a constant of integration show up on the left side after doing the integration. We can drop it at this point since other constants of integration will be showing up down the road and they would just end up absorbing this one. Finally, rewrite the formula as follows and we arrive that the integration by parts formula.

$$\int fg'\,dx = fg - \int f'g\,dx$$

This is not the easiest formula to use however. So, let's do a couple of substitutions.

$$u = f(x) \qquad v = g(x)$$
$$du = f'(x)\,dx \qquad dv = g'(x)\,dx$$

Both of these are just the standard Calculus substitutions that hopefully you are used to by now. Don't get excited by the fact that we are using two substitutions here. They will work the same way. Using these substitutions gives us the formula that most people think of as the integration by parts formula.

$$\int u\,dv = uv - \int v\,du$$

To use this formula we will need to identify u and dv, compute du and v and then use the formula. Note as well that computing v is very easy. All we need to do is integrate dv.

$$v = \int dv$$

So, let's take a look at the integral above that we mentioned we wanted to do.

Example: Evaluate the following integral.

$$\int xe^{6x}\,dx$$

Solution

So, on some level, the problem here is the x that is in front of the exponential. If that wasn't there we could do the integral. Notice as well that in doing integration by parts anything that we choose for u will be differentiated. So, it seems that choosing $u = x$ will be a good choice since upon differentiating the x will drop out. Now that we've chosen u we know that dv will be everything else that remains. So, here are the choices for u and dv as well as du and v.

$$u - v \qquad dv = e^{6x}\,dx$$

$$du = dx \qquad v = \int e^{6x}\,dx = \frac{1}{2}e^{6x}$$

The integral is then,

$$\int xe^{6x}\,dx = \frac{x}{6}e^{6x} - \int \frac{1}{6}e^{6x}\,dx$$

$$= \frac{x}{6}e^{6x} - \frac{1}{36}e^{6x} + c$$

Once we have done the last integral in the problem we will add in the constant of integration to get our final answer. Next, let's take a look at integration by parts for definite integrals. The integration by parts formula for definite integrals is,

DEFINITE INTEGRALS

$$\int_a^b u\,dv = uv\Big|_a^b - \int_a^b v\,du$$

Note that the $uv\Big|_a^b$ in the first term is just the standard integral evaluation notation that you should be familiar with at this point. All we do is evaluate the term, uv in this case, at b then subtract off the evaluation of the term at a. At some level we don't really need a formula here because we know that when doing definite integrals all we need to do is do the indefinite integral and then do the evaluation. Let's take a quick look at a definite integral using integration by parts.

Example: Evaluate the following integral.

$$\int_{-1}^{2} xe^{6x}\,dx$$

Solution: This is the same integral that we looked at in the first example so we'll use the same u and dv to get,

$$\begin{aligned}\int_{-1}^{2} xe^{6x}\,dx &= \frac{x}{6}e^{6x}\Big|_{-1}^{2} - \frac{1}{6}\int_{-1}^{2} e^{6x}\,dx \\ &= \frac{x}{6}e^{6x}\Big|_{-1}^{2} - \frac{1}{36}e^{6x}\Big|_{-1}^{2} \\ &= \frac{11}{36}e^{12} + \frac{7}{36}e^{-6}\end{aligned}$$

Since we need to be able to do the indefinite integral in order to do the definite integral and doing the definite integral amounts to nothing more than evaluating the indefinite integral at a couple of points we will concentrate on doing indefinite integrals in the rest of this section. In fact, throughout most of this chapter this will be the case.

We will be doing far more indefinite integrals than definite integrals.

Let's take a look at some more examples.

Example: Evaluate the following integral.

$$\int (3t+5)\cos\left(\frac{1}{4}\right)dt$$

Solution: There are two ways to proceed with this example. For many, the first thing that they try is multiplying the cosine through the parenthesis, splitting up the integral and then doing integration by parts on the first integral.

While that is a perfectly acceptable way of doing the problem it's more work than we really need to do. Instead of splitting the integral up let's instead use the following choices for u and dv.

$$u = 3t+5 \qquad dv = \cos\left(\frac{1}{4}\right)dt$$

$$du = 3dt \qquad v = 4\sin\left(\frac{t}{4}\right)$$

The integral is then,

$$\int (3t+5)\cos\left(\frac{t}{4}\right)dt = 4(3t+5)\sin\left(\frac{t}{4}\right) - 12\int \sin\left(\frac{1}{4}\right)dt$$

$$= 4(3t+5)\sin\left(\frac{t}{4}\right) + 48\cos\left(\frac{t}{4}\right) + c$$

Notice that we pulled any constants out of the integral when we used the integration by parts formula. We will usually do this in order to simplify the integral a little.

Example: Evaluate the following integral.

$$\int w^2 \sin(10w)\,dw$$

Solution: For this example we'll use the following choices for u and dv.

$$u = w^2 \qquad dv = \sin(10w)\,dw$$

$$du = 2w\,dw \qquad v = -\frac{1}{10}\cos(10w)$$

The integral is then,

$$\int w^2 \sin(10w)\,dw = -\frac{w^2}{10}\cos(10w) + \frac{1}{5}\int w\cos(10w)\,dw$$

In this example, unlike the previous examples, the new integral will also require integration by parts. For this second integral we will use the following choices.

$$u = w \qquad dv = \cos(10w)\,dw$$

$$du = dw \qquad v = \frac{1}{10}\sin(10w)$$

So, the integral becomes,

$$\begin{aligned}\int w^2 \sin(10w)\,dw &= -\frac{w^2}{10}\cos(10w) + \frac{1}{5}\left(\frac{w}{10}\sin(10w) - \frac{1}{10}\int \sin(10w)\,dw\right)\\ &= -\frac{w^2}{10}\cos(10w) + \frac{1}{5}\left(\frac{w}{10}\sin(10w) - \frac{1}{100}\cos(10w)\right) + c\\ &= -\frac{w^2}{10}\cos(10w) + \frac{w}{50}\sin(10w) + \frac{1}{500}\cos(10w) + c\end{aligned}$$

Be careful with the coefficient on the integral for the second application of integration by parts. Since the integral is multiplied by we need to make sure that the results of actually doing the integral are also multiplied by. Forgetting to do this is one of the more common mistakes with integration by parts problems. As this last example has shown us, we will sometimes need more than one application of integration by parts to completely evaluate an integral. This is something that will happen so don't get excited about it when it does.

In this next example we need to acknowledge an important point about integration techniques. Some integrals can be done in using several different techniques. That is the case with the integral in the next example.

Example: Evaluate the following integral

$$\int x\sqrt{x+1}\,dx$$

1. Using Integration by Parts.
2. Using a standard Calculus substitution.

Solution:

1. Evaluate using Integration by Parts.

First notice that there are no trig functions or exponentials in this integral. While a good many integration by parts integrals will involve trig functions and/or exponentials not all of them will so don't get too locked into the idea of expecting them to show up. In this case we'll use the following choices for u and dv.

$$u = x \qquad dv = \sqrt{x+1}\,dx$$

$$du = dx \qquad v = \frac{2}{3}(x+1)^{\frac{3}{2}}$$

The integral is then,

$$\int x\sqrt{x+1}\,dx = \frac{2}{3}x(x-1)^{\frac{3}{2}} - \frac{2}{3}\int (x+1)^{\frac{3}{2}}\,dx$$

$$= \frac{2}{3}x(x-1)^{\frac{3}{2}} - \frac{4}{15}(x+1)^{\frac{5}{2}} + c$$

2. Evaluate Using a standard Calculus substitution.

Now let's do the integral with a substitution. We can use the following substitution.

$$u = x+1 \qquad x = u-1 \qquad du = dx$$

Notice that we'll actually use the substitution twice, once for the quantity under the square root and once for the x in front of the square root. The integral is then,

$$\int x\sqrt{x+1}\,dx = \int (u-1)\sqrt{u}\,du$$

$$= \int u^{\frac{3}{2}} - u^{\frac{1}{2}}\,du$$

$$= \frac{2}{5}u^{\frac{5}{2}} - \frac{2}{3}u^{\frac{3}{2}} + c$$

$$= \frac{2}{5}(x+1)^{\frac{5}{2}} - \frac{2}{3}(x+1)^{\frac{3}{2}} + c$$

So, we used two different integration techniques in this example and we got two different answers.

The obvious question then should be: Did we do something

wrong? Actually, we didn't do anything wrong. We need to remember the following fact from Calculus.

$$\text{if } f'(x)=g'(x) \text{ then } f(x)\;\; g(x)+c$$

In other words, if two functions have the same derivative then they will differ by no more than a constant. So, how does this apply to the above problem? First define the following,

$$f'(x)=g'(x)=x\sqrt{x+1}$$

Then we can compute *f(x)* and *g(x)* by integrating as follows,

$$f(x)=\int f'(x)\,dx \qquad g(x)=\int g'(x)\,dx$$

We'll use integration by parts for the first integral and the substitution for the second integral. Then according to the fact *f(x)* and *g(x)* should differ by no more than a constant. Let's verify this and see if this is the case. We can verify that they differ by no more than a constant if we take a look at the difference of the two and do a little algebraic manipulation and simplification.

$$\begin{aligned}&\left(\frac{2}{3}x(x+1)^{\frac{3}{2}}-\frac{4}{15}(x+1)^{\frac{5}{2}}\right)-\left(\frac{2}{5}(x+1)^{\frac{5}{2}}-\frac{2}{5}(x+1)^{\frac{3}{2}}\right)\\&=(x+1)^{\frac{3}{2}}\left(\frac{2}{3}x-\frac{4}{15}(x+1)-\frac{2}{5}(x+1)+\frac{2}{5}\right)\\&=(x+1)^{\frac{3}{2}}(0)\\&=0\end{aligned}$$

So, in this case it turns out the two functions are exactly the same function since the difference is zero. Note that this won't always happen. Sometimes the difference will yield a nonzero constant. For an example of this check out the Constant of Integration section in my Calculus notes. So just what have we learned? First, there will, on occasion, be more than one method for evaluating an integral. Secondly, we saw that different methods will often lead to different answers. Last, even though the answers are different it can be shown, sometimes with a lot of work, that they differ by no more than a constant.

When we are faced with an integral the first thing that we'll need to decide is if there is more than one way to do the integral. If there is more than one way we'll then need to determine which method we should use. The general rule of thumb that I use in my classes is that you should use the method that *you* find easiest. This may not be the method that others find easiest, but that doesn't make it the wrong method.

One of the more common mistakes with integration by parts is for people to get too locked into perceived patterns. For instance, all of the previous examples used the basic pattern of taking u to be the polynomial that sat in front of another function and then letting dv be the other function. This will not always happen so we need to be careful and not get locked into any patterns that we think we see. Let's take a look at some integrals that don't fit into the above pattern.

Example: Evaluate the following integral.

$$\int In\, xdx$$

Solution: So, unlike any of the other integral we've done to this point there is only a single function in the integral and no polynomial sitting in front of the logarithm. The first choice of many people here is to try and fit this into the pattern from above and make the following choices for u and dv.

$$u = 1 \qquad\qquad dv = In\, xdx$$

This leads to a real problem however since that means v must be,

$$v = \int \text{In}\, xdx$$

In other words, we would need to know the answer ahead of time in order to actually do the problem. So, this choice simply won't work. Also notice that with this choice we'd get that which also causes problems and is another reason why this choice will not work. Therefore, if the logarithm doesn't belong in the dv it must belong instead in the u. So, let's use the following choices instead

$$u = \text{In}\, x \qquad\qquad dv = dx$$

$$du = \frac{1}{2}dx \qquad\qquad v = x$$

The integral is then,

$$\int \ln x\,dx = x\ln x - \int \frac{1}{x}x\,dx$$

$$= x\ln x - \int dx$$

$$= x\ln x - x + c$$

Example: Evaluate the following integral.

$$\int x^5\sqrt{x^2+1dx}$$

Solution: So, if we again try to use the pattern from the first few examples for this integral our choices for u and dv would probably be the following.

$$u = x^5 \qquad dv = \sqrt{x^3+1dx}$$

However, as with the previous example this won't work since we can't easily compute v.

$$v = \int \sqrt{x^3+1dx}$$

This is not an easy integral to do. However, notice that if we had an x^2 in the integral along with the root we could very easily do the integral with a substitution. Also notice that we do have a lot of x's floating around in the original integral. So instead of putting all the x's (outside of the root) in the u let's split them up as follows.

$$u = x^3 \qquad dv = x^2\sqrt{x^3+1}\,dx$$

$$du = 3x^2dx \qquad v = \frac{2}{9}(x^3+1)^{\frac{3}{2}}$$

We can now easily compute v and after using integration by parts we get,

$$\int x^5\sqrt{x^3+1}\,dx = \frac{2}{9}x^3(x^3+1)^{\frac{3}{2}} - \frac{2}{3}\int x^2(x^3+1)^{\frac{3}{2}}\,dx$$

$$= \frac{2}{9}x^3(x^3+1)^{\frac{3}{2}} - \frac{4}{45}(x^3-1)^{\frac{3}{2}} + c$$

So, in the previous two examples we saw cases that didn't

quite fit into any perceived pattern that we might have gotten from the first couple of examples. This is always something that we need to be on the lookout for with integration by parts. Let's take a look at another example that also illustrates another integration technique that sometimes arises out of integration by parts problems.

Example: Evaluate the following integral.

$$\int e^{\theta} \cos\theta \, d\theta$$

Solution: Okay, to this point we've always picked u in such a way that upon differentiating it would make that portion go away or at the very least put it the integral into a form that would make it easier to deal with.

In this case no matter which part we make u it will never go away in the differentiation process. It doesn't much matter which we choose to be u so we'll choose in the following way. Note however that we could choose the other way as well and we'll get the same result in the end.

$$u = \cos\theta$$
$$dv = e^{\theta} \, d\theta$$
$$du = -\sin\theta \, d\theta \qquad v = e^{\theta}$$

The integral is then,

$$\int e^{\theta} \cos\theta \, d\theta = e^{\theta} \cos\theta + \int e^{\theta} \sin\theta \, d\theta$$

So, it looks like we'll do integration by parts again. Here are our choices this time.

$$u = \sin\theta \qquad dv = e^{\theta} d\theta$$

$$du = \cos\theta \, d\theta \qquad v = e^{\theta}$$

The integral is now,

$$\int e^{\theta} \cos\theta \, d\theta = e^{\theta} \cos\theta + e^{\theta} \sin\theta - \int e^{\theta} \cos\theta \, d\theta$$

Now, at this point it looks like we're just running in circles. However, notice that we now have the same integral on both sides and on the right side it's got a minus sign in front of it. This means that we can add the integral to both sides to get,

$$2\int e^{\theta}\cos\theta\, d\theta = e^{\theta}\cos\theta + e^{\theta}\sin\theta$$

All we need to do now is divide by 2 and we're done. The integral is,

$$\int e^{\theta}\cos\theta\, d\theta = \frac{1}{2}\left(e^{\theta}\cos\theta + e^{\theta}\sin\theta\right) + c$$

Notice that after dividing by the two we add in the constant of integration at that point. This idea of integrating until you get the same integral on both sides of the equal sign and then simply solving for the integral is kind of nice to remember. It doesn't show up all that often, but when it does it may be the only way to actually do the integral. We've got one more example to do. As we will see some problems could require us to do integration by parts numerous times and there is a short hand method that will allow us to do multiple applications of integration by parts quickly and easily.

Example: Evaluate the following integral.

$$\int x^4 e^{\frac{x}{2}}\, dx$$

Solution: We start off by choosing *u* and *dv* as we always would. However, instead of computing *du* and *v* we put these into the following table. We then differentiate down the column corresponding to *u* until we hit zero. In the column corresponding to *dv* we integrate once for each entry in the first column. There is also a third column which we will explain in a bit and it always starts with a "+" and then alternates signs as shown. Now, multiply along the diagonals show in the table. In front of each product put the sign in the third column that corresponds to the "*u*" term for that product. In this case this would give,

$$\int x^4 e^{\frac{x}{2}}\, dx = \left(x^4\right)(2e^{\frac{x}{2}}) - (4x^3)(4e^{x}2) + (12x^2)(8e^{\frac{x}{2}})$$

$$-(24x)(16e^{\frac{x}{2}}) \cdot (24)(32e^{\frac{x}{2}})$$

$$= 2x^4 e^{\frac{x}{2}} \quad 16x^3 e^{\frac{x}{2}} \quad 96x^2 e^{\frac{x}{2}} \quad 384xe^{\frac{x}{2}} \cdot 786\, e^{\frac{x}{2}} \mid c$$

x^4	$e^{\frac{x}{2}}$	+
$4x^3$	$2e^{\frac{x}{2}}$	−
$12x^2$	$4e^{\frac{x}{2}}$	+
$24x$	$8e^{\frac{x}{2}}$	−
24	$16e^{\frac{x}{2}}$	+
0	$32e^{\frac{x}{2}}$	−

We've got the integral. This is much easier than writing down all the various u's and dv's that we'd have to do otherwise.

INTEGRALS INVOLVING TRIG FUNCTIONS

In this section we are going to look at quite a few integrals involving trig functions and some of the techniques we can use to help us evaluate them. Let's start off with an integral that we should already be able to do.

$$\int \cos x \sin^5 x\,dx = \int u^5\,du \quad \text{using the substitution } u = \sin x$$

$$= \frac{1}{6}\sin^6 x + c$$

This integral is easy to do with a substitution because the presence of the cosine, however, what about the following integral.

Example: Evaluate the following integral.

$$\int \sin^5 x dx$$

Solution: This integral no longer has the cosine in it that would allow us to use the substitution that we used above. Therefore, that substitution won't work and we are going to have to find another way of doing this integral.

Let's first notice that we could write the integral as follows,

$$\int \sin^5 x dx = \int \sin^4 x \sin x dx = \int \left(\sin^2 x\right)^2 \sin x dx$$

Now recall the trig identity,

$$\cos^2 x + \sin^2 x + 1 \qquad \Rightarrow \qquad \sin^2 x + 1 - \cos^2 x$$

With this identity the integral can be written as,

$$\int \sin^5 x dx = \int \left(1 - \cos^2 x\right)^2 \sin x dx$$

and we can now use the substitution *u* = *cos x*. Doing this gives us,

$$\int \sin^5 x\, dx = \int (1 - u^2)^2 du$$

$$= -\int 1 - 2u^2 + u^4 du$$

$$= -\left(u - \frac{2}{3}u^3 + \frac{1}{5}u^5 \right) + c$$

$$= -\cos x + \frac{2}{3}\cos^3 x - \frac{1}{5}\cos^5 x + c$$

So, with a little rewriting on the integrand we were able to reduce this to a fairly simple substitution.

Notice that we were able to do the rewrite that we did in the previous example because the exponent on the sine was odd. In these cases all that we need to do is strip out one of the sines. The exponent on the remaining sines will then be even and we can easily convert the remaining sines to cosines using the identity,

$$\cos^2 x + \sin^2 x = 1$$

If the exponent on the sines had been even this would have been difficult to do.

We could strip out a sine, but the remaining sines would then have an odd exponent and while we could convert them to cosines the resulting integral would often be even more difficult than the original integral in most cases.

Let's take a look at another example.

Example: Evaluate the following integral.

$$\int \sin^6 x \cos^3 x dx$$

Solution: So, in this case we've got both sines and cosines in the problem and in this case the exponent on the sine is even while the exponent on the cosine is odd. So, we can use a similar technique in this integral. This time we'll strip out a cosine and convert the rest to sines.

$$\begin{aligned}\int \sin^6 x \cos^3 x\,dx &= \int \sin^6 x \cos^2 x \cos x\,dx \\ &= \int \sin^6 x(1-\sin^2 x)\cos x\,dx \qquad u = \sin x \\ &= -\left(u - \frac{2}{3}u^3 + \frac{1}{5}u^5\right) + c \\ &= \int u^6(1-u^2)\,dx \\ &= \int u^6 - u^8 du \\ &= \frac{1}{7}\sin^7 x - \frac{1}{9}\sin^9 x + c\end{aligned}$$

At this point let's pause for a second to summarize what we've learned so far about integrating powers of sine and cosine.

$$\int \sin^n x \cos^m x dx$$

In this integral if the exponent on the sines (n) is odd we can strip out one sine, convert the rest to cosines using and then use the substitution.

Likewise, if the exponent on the cosines (m) is odd we can strip out one cosine and convert the rest to sines and the use the substitution $u = sin\ x$.

Of, course if both exponents are odd then we can use either method. However, in these cases it's usually easier to convert the term with the smaller exponent.

The one case we haven't looked at is what happens if both of the exponents are even? In this case the technique we used in the first couple of examples simply won't work and in fact there really isn't any one set method for doing these integrals.

Each integral is different and in some cases there will be more than one way to do the integral. With that being said most, if not all, of integrals involving products of sines and cosines in which both exponents are even can be done using one or more of the following formulas to rewrite the integrand.

$$\sin^2 x = \frac{1}{2}(1 + \sin(2x))$$

$$\cos^2 x = \frac{1}{2}(1 + \cos(2x))$$

$$\sin x \cos x = \frac{1}{2}\sin(2x))$$

The first two formulas are the standard half angle formula from a trig class written in a form that will be more convenient for us to use. The last is the standard double angle formula for sine, again with a small rewrite.

Let's take a look at an example.

Example: Evaluate the following integral.

$$\int \sin^2 \cos^2 x dx$$

Solution: As noted above there are often more than one way to do integrals in which both of the exponents are even. This integral is an example of that. There are at least two solution techniques for this problem. We will do both solutions starting with what is probably the harder of the two, but it's also the one that many people see first. In this solution we will use the two half angle formulas above and just substitute them into the integral.

$$\int \sin^2 \cos^2 x dx = \int \frac{1}{2}\left(1 - \cos\left(2x\right)\right)\left(\frac{1}{2}\right)\left(1 + \cos\left(2x\right)\right)dx$$

$$= \frac{1}{4}\int 1 - \cos^2\left(2x\right)dx$$

So, we still have an integral that can't be completely done, however notice that we have managed to reduce the integral down to just one term causing problems (a cosine with an even power) rather than two terms causing problems. In fact to eliminate the remaining problem term all that we need to do is reuse the first half angle formula given above.

$$\int \sin^2 x \cos^2 x\, dx = \frac{1}{4}\int 1 - \frac{1}{2}(1 + \cos(4x))dx$$

$$= \frac{1}{4}\int \frac{1}{2} - \frac{1}{2}\cos(4x))dx$$

$$=\frac{1}{4}\left(\frac{1}{2}x-\frac{1}{8}\sin(4x)\right)+c$$

$$=\frac{1}{8}x-\frac{1}{32}\sin(4x)+c$$

So, this solution required a total of three trig identities to complete.

Solution: In this solution we will use the half angle formula to help simplify the integral as follows.

$$\int\sin^2 x\cos^2 x\,dx=\int(\sin x\cos x)^2\,dx$$

$$=\int\left(\frac{1}{2}\sin(2x)\right)^2 dx$$

$$=\frac{1}{4}\int\sin^2(2x)\,dx$$

Now, we use the double angle formula for sine to reduce to an integral that we can do.

$$\int\sin^2 x\cos^2 x\,dx=\frac{1}{8}\int 1-\cos(4x)\,dx$$

$$=\frac{1}{8}x+\frac{1}{32}\sin(4x)+c$$

This method required only two trig identities to complete. Notice that the difference between these two methods is more one of "messiness". The second method is not appreciably easier (other than needing one less trig identity) it is just not as messy and that will often translate into an "easier" process.

In the previous example we saw two different solution methods that gave the same answer. Note that this will not always happen. In fact, more often than not we will get different answers. However, as we discussed in the Integration by Parts section, the two answers will differ by no more than a constant. In general when we have products of sines and cosines in which both exponents are even we will need to use a series of half angle and/

or double angle formulas to reduce the integral into a form that we can integrate.

Also, the larger the exponents the more we'll need to use these formulas and hence the messier the problem. Sometimes in the process of reducing integrals in which both exponents are even we will run across products of sine and cosine in which the arguments are different.

These will require one of the following formulas to reduce the products to integrals that we can do.

$$\sin\alpha\cos\beta = \frac{1}{2}\left[\sin(\alpha+\beta)+\sin(\alpha+\beta)\right]$$

$$\sin\alpha\sin\beta = \frac{1}{2}\left[\sin(\alpha-\beta)-\cos(\alpha+\beta)\right]$$

$$\cos\alpha\cos\beta = \frac{1}{2}\left[\cos(\alpha-\beta)+\cos(\alpha-\beta)\right]$$

Let's take a look at an example of one of these kinds of integrals.

Example: Evaluate the following integral.

$$\int \cos(15x)\cos(4x)\,dx$$

Solution: This integral requires the last formula listed above.

$$\int \cos(15x)\cos(4x)\,dx = \frac{1}{2}\int \cos(11x)+\cos(19x)\,dx$$

$$= \frac{1}{2}\left(\frac{1}{11}\sin(11x)+\frac{1}{19}\sin(19x)\right)+c$$

Okay, at this point we've covered pretty much all the possible cases involving products of sines and cosines.

It's now time to look at integrals that involve products of secants and tangents. This time, let's do a little analysis of the possibilities before we just jump into examples. The general integral will be,

$$\int \sec^n x\tan^m x\,dx$$

The first thing to notice is that we can easily convert even powers of secants to tangents and even powers of tangents to secants by using a formula. In fact, the formula can be derived so let's do that.

$$\sin^2 x + \cos^2 x = 1$$

$$\frac{\sin^2 x}{\cos^2 x} + \frac{\cos^2 x}{\cos^2 x} = \frac{1}{\cos^2 x} \qquad \tan^2 x + 1 = \sec^2 x$$

We'll want to eventually use one of the following substitutions.

$$u = \tan x \qquad du = \sec^2 x dx$$

$$u = \sec x \qquad du = \sec x \tan x dx$$

So, if we use the substitution we will need two secants left for the substitution to work. This means that if the exponent on the secant (n) is even we can strip two out and then convert the remaining secants to tangents.

Next, if we want to use the substitution $u = sec\ x$ we will need one secant and one tangent left over in order to use the substitution. This means that if the exponent on the tangent (m) is odd and we have at least one secant in the integrand we can strip out one of the tangents along with one of the secants of course.

The tangent will then have an even exponent to convert the rest to tangents to secants. Note that this method does require that we have at least one secant in the integral as well. If there aren't any secants then we'll need to do something different. If the exponent on the secant is even and the exponent on the tangent is odd then we can use either case. Again, it will be easier to convert the term with the smallest exponent.

Let's take a look at a couple of examples.

Example: Evaluate the following integral.

$$\int \sec^9 x \tan^5 x dx$$

Solution: First note that since the exponent on the secant isn't even we can't use the substitution. However, the exponent on the tangent is odd and we've got a secant in the integral and so we will be able to use the substitution $u = sec\ x$. This means striping out a single tangent (along with a secant) and converting the remaining tangents to secants. Here's the work for this integral.

$$\begin{aligned}\int \sec^9 x \tan^5 x\,dx &= \int \sec^8 x \tan^4 x \tan x \sec x\,dx \\ &= \int \sec^8 x(\sec^2 x - 1)^2 \tan x \sec x\,dx \qquad u = \sec x \\ &= \int u^8(u^2-1)^2\,du \\ &= \int u^{12} - 2u^{10} + u^8\,du\end{aligned}$$

Example: Evaluate the following integral.

$$\int \sec^4 x \tan^6 x dx$$

Solution: So, in this example the exponent on the tangent is even so the substitution won't work. The exponent on the secant is even and so we can use the substitution for this integral. That means that we need to strip out two secants and convert the rest to tangents. Here is the work for this integral.

$$\begin{aligned}\int \sec^4 x \tan^6 x\,dx &= \int \sec^2 x \tan^6 x \sec^2 x\,dx \\ &= \int (\tan^2 x + 1)\tan^6 x \sec^2 x\,dx \qquad u = \tan x \\ &= \int (u^2-1)^2 u^6 du \\ &= \int u^8 + u^6\,du \\ &= \frac{1}{9}\tan^9 x + \frac{1}{7}\tan^7 x + c\end{aligned}$$

Both of the previous examples fit very nicely into the patterns discussed above and so were not all that difficult to work. However, there are a couple of exceptions to the patterns above and in these cases there is no single method that will work for every problem. Each integral will be different and may require different solution methods in order to evaluate the integral.

Let's first take a look at a couple of integrals that have odd exponents on the tangents, but no secants. In these cases we can't use the substitution $u = sec\ x$ since it requires there to be at least one secant in the integral.

Example: Evaluate the following integral.

$$\int \tan x\, dx$$

Solution: To do this integral all we need to do is recall the definition of tangent in terms of sine and cosine and then this integral is nothing more than a Calculus substitution.

$$\begin{aligned}\int \tan x\, dx &= \int \frac{\sin x}{\cos x} dx = \cos x\\ &= -\int \frac{1}{u} du\\ &= -\ln|\cos x| + c \qquad\qquad r\ln x = \ln x^{r}\\ &= \ln|\cos x|^{1} + c\\ &= \ln|\sec x| + c\end{aligned}$$

Example: Evaluate the following integral.

$$\int \tan^3 x dx$$

Solution: The trick to this one is do the following manipulation of the integrand.

$$\begin{aligned}\int \tan^3 x dx &= \int \tan x \tan^2 x dx\\ &= \int \tan x\left(\sec^2 x + 1\right)\\ &= \int \tan x \sec^2 x dx - \int \tan x dx\end{aligned}$$

We can now use the substitution on the first integral and the results from the previous example to on the second integral.

The integral is then, $\int \tan^3 x dx = \frac{1}{2}\tan^2 x - In|\sec x| + c$

Note that all odd powers of tangent (with the exception of

the first power) can be integrated using the same method we used in the previous example. For instance,

$$\int \tan^2 x dx = \int \tan^3 x\left(\sec^2 x - 1\right) dx = \int \tan^3 x \sec^2 x dx - \int \tan^3 x dx$$

So, a quick substitution (u = *tan x*) will give us the first integral and the second integral will always be the previous odd power. Now let's take a look at a couple of examples in which the exponent on the secant is odd and the exponent on the tangent is even. In these cases the substitutions used above won't work. It should also be noted that both of the following two integrals are integrals that we'll be seeing on occasion in later sections of this chapter and in later chapters. Because of this it wouldn't be a bad idea to make a note of these results so you'll have them ready when you need them later.

Example: Evaluate the following integral.

$$\int \sec x dx$$

Solution: This one isn't too bad once you see what you've got to do. By itself the integral can't be done. However, if we manipulate the integrand as follows we can do it.

$$\int \sec x\, dx = \int \frac{\sec x(\sec x + \tan x)}{\sec x + \tan x} dx$$

$$= \int \frac{\sec^2 x \tan x \sec x}{\sec x + \tan x} dx$$

In this form we can do the integral using the substitution u = *sec x* + *tan x*. Doing this gives,

$$\int \sec x dx = \text{In}\left|\sec \text{x} \tan \text{x}\right| + c$$

The idea used in the above example is a nice idea to keep in mind. Multiplying the numerator and denominator of a term by the same term above can, on occasion, put the integral into a form that can be integrated. Note that this method won't always work and even when it does it won't always be clear what you need to multiply the numerator and denominator by. However, when it does work and you can figure out what term you need it can greatly simplify the integral.

Here's the next example.

Example: Evaluate the following integral.

$$\int \sec^3 x dx$$

Solution: This one is different from any of the other integrals that we've done in this section. The first step to doing this integral is to perform integration by parts using the following choices for u and dv.

$$u = \sec x \qquad dv = \sec^2 x dx$$

$$du = \sec x \tan x dx \qquad v = \tan x$$

Note that using integration by parts on this problem is not an obvious choice, but it does work very nicely here. After doing integration by parts we have,

$$\int \sec^2 x dx = \sec x \tan x - \int \sec x \tan^2 x dx$$

Now the new integral also has an odd exponent on the secant and an even exponent on the tangent and so the previous examples of products of secants and tangents still won't do us any good. To do this integral we'll first write the tangents in the integral in terms of secants.

Again, this is not necessarily an obvious choice but it's what we need to do in this case.

$$\int \sec^2 x dx = \sec x \tan x - \int \sec x \left(\sec^2 x - 1\right) dx$$

$$= \sec x \tan x - \int \sec^3 x dx + \int \sec x dx$$

Now, we can use the results from the previous example to do the second integral and notice that the first integral is exactly the integral we're being asked to evaluate with a minus sign in front. So, add it to both sides to get,

$$2\int \sec^2 x dx = \sec x \tan x + \text{In}\left|\sec x + \tan x\right|$$

Finally divide by a two and we're done.

$$\int \sec^3 x dx = \frac{1}{2}\left(\sec x \tan x + \text{In}\left|\sec x + \tan x\right|\right) + c$$

Again, note that we've again used the idea of integrating the right side until the original integral shows up and then moving this to the left side and dividing by its coefficient to complete the evaluation.

We first saw this in the Integration by Parts section and noted at the time that this was a nice technique to remember. Here is another example of this technique. Now that we've looked at products of secants and tangents let's also acknowledge that because we can relate cosecants and cotangents by

$$1 + \cot^2 x = \csc^2 x$$

all of the work that we did for products of secants and tangents will also work for products of cosecants and cotangents. We'll leave it to you to verify that. There is one final topic to be discussed in this section before moving on.

To this point we've looked only at products of sines and cosines and products of secants and tangents. However, the methods used to do these integrals can also be used on some quotients involving sines and cosines and quotients involving secants and tangents (and hence quotients involving cosecants and cotangents). Let's take a quick look at an example of this.

Example: Evaluate the following integral.

$$\int \frac{\sin^7 x}{\cos^4 x} dx$$

Solution: If this were a product of sines and cosines we would know what to do. We would strip out a sine (since the exponent on the sine is odd) and convert the rest of the sines to cosines.

The same idea will work in this case. We'll strip out a sine from the numerator and convert the rest to cosines as follows,

$$\int \frac{\sin^7 x}{\cos^4 x} dx = \int \frac{\sin^6 x}{\cos^4 x} \sin x \, dx$$

$$= \int \frac{(\sin^2 x)^3}{\cos^4 x} \sin x\, dx$$

$$= \int \frac{(1-\cos^2 x)^3}{\cos^4 x} \sin x\, dx$$

At this point all we need to do is use the substitution and we're done.

$$\int \frac{\sin^7 x}{\cos^4 x} dx = \int \frac{(1-u^2)^3}{u^4} du$$

$$= -\int u^{-4} - 3u^{-2} + 3 - u^2 du$$

$$= -\left(-\frac{1}{3}\frac{1}{u^3} + 3\frac{1}{u} + 3u - \frac{1}{3}u^3\right) + c$$

$$= \frac{1}{3\cos^3 x} - \frac{3}{\cos x} - 3\cos x + \frac{1}{3}\cos^3 x + c$$

TRIG SUBSTITUTIONS

As we have done in the last couple of sections, let's start off with a couple of integrals that we should already be able to do with a standard substitution.

$$\int x\sqrt{25x^2 - 4}\, dx = \frac{1}{75}(25x^2 - 4)^{\frac{3}{2}} + c$$

$$\int \frac{x}{\sqrt{25x^2 - 4}} dx = \frac{1}{25}\sqrt{25x^2 - 4} + c$$

Both of these used the substitution $u = 25x^2 - 4$ and at this point should be pretty easy for you to do. However, let's take a look at the following integral.

Example: Evaluate the following integral.

$$\int \frac{\sqrt{25x^2 - 4}}{x} dx$$

Solution: In this case the substitution $u = 25x^2 - 4$ will not work and so we're going to have to do something different for this integral. It would be nice if we could get rid of the square root somehow. The following substitution will do that for us.

$$x = \frac{2}{5}\sec\theta$$

Do not worry about where this came from at this point. As we work the problem you will see that it works and that if we have a similar type of square root in the problem we can always use a similar substitution. Before we actually do the substitution however let's verify the claim that this will allow us to get rid of the square root.

$$\sqrt{25x^2 - 4} = \sqrt{25\left(\frac{4}{25}\right)\sec^2\theta - 4} = \sqrt{4(\sec^2\theta - 1)} = 2\sqrt{\sec^2\theta - 1}$$

To get rid of the square root all we need to do is recall the relationship,

$$\tan^2\theta + 1 = \sec^2\theta \qquad \Rightarrow \qquad \sec^2\theta - 1 = \tan^2\theta$$

Using this fact the square root becomes,

$$\sqrt{25x^2 - 4} = 2\sqrt{\tan^2\theta} = 2\left|\tan\theta\right|$$

Note the presence of the absolute value bars there. These are important. Recall that

$$\sqrt{x^2} = |x|$$

There should always be absolute value bars at this stage. If we knew that $tan\theta$ was always positive or always negative we could eliminate the absolute value bars using,

$$|x| = \begin{cases} x & \text{if } x \geq 0 \\ -x & \text{if } x \geq 0 \end{cases}$$

Without limits we won't be able to determine if $tan\ \theta$ is positive of negative, however, we will need to eliminate them in order to do the integral. Therefore, since we doing an indefinite integral we will assume that $tan\ \theta$ will be positive and so we can drop the absolute value bars. This gives, $\sqrt{25x^2 - 4} = 2\tan\theta$

So, we were able to eliminate the square root using this substitution. Let's now do the substitution and see what we get.

In doing the substitution don't forget that well also need to substitute for the dx. This is easy enough to get from the substitution.

$$x = \frac{2}{5}\sec\theta \qquad \Rightarrow \qquad dx = \frac{2}{5}\sec\theta\tan\theta d\theta$$

Using this substitution the integral becomes,

$$\int \frac{\sqrt{25x^2-4}}{x}dx = \int \frac{2\tan\theta}{\frac{2}{5}\sec\theta}\left(\frac{2}{5}\sec\theta\tan\theta\right)d\theta$$

$$= 2\int \tan^2\theta d\theta$$

With this substitution we were able to reduce the given integral to an integral involving trig functions and we saw how to do these problems in the previous section. Let's finish the integral.

$$\int \frac{\sqrt{25x^2-4}}{x}dx = 2\int \sec^2\theta - 1 d\theta$$

$$= 2(\tan\theta - \theta) + c$$

So, we've got an answer for the integral. Unfortunately the answer isn't given in x's as it should be. So, we need to write our answer in terms of x. We can do this with some right triangle trig. From our original substitution we have,

$$\sec\theta = \frac{5x}{2} = \frac{\text{hypotenuse}}{\text{adjacent}}$$

This gives the following right triangle.

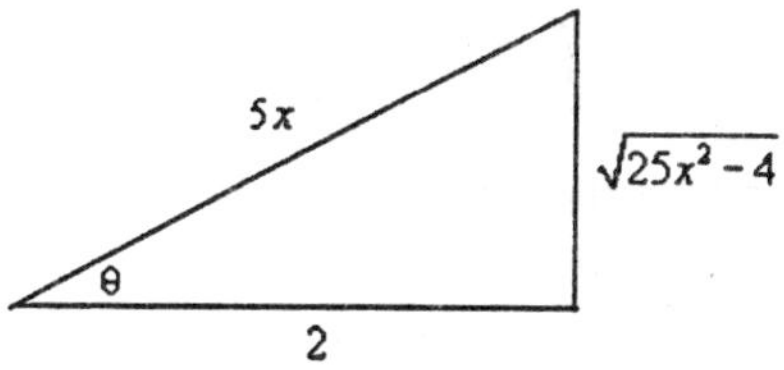

From this we can see that,

$$\tan\theta = \frac{\sqrt{25x^2-4}}{2}$$

We can deal with the in one of any variety of ways. From our substitution we can see that,

$$\theta = \sec^{-1}\left(\frac{5x}{2}\right)$$

While this is a perfectly acceptable method of dealing with the we can use any of the possible six inverse trig functions and since sine and cosine are the two trig functions most people are familiar with we will usually use the inverse sine or inverse cosine. In this case we'll use the inverse cosine.

$$\theta = \cos^{-1}\left(\frac{2}{5x}\right)$$

So, with all of this the integral becomes,

$$\int \frac{\sqrt{25x^2-4}}{x}dx = 2\left(\frac{\sqrt{25x^2-4}}{x} - \cos^{-2}\left(\frac{2}{5x}\right)\right) + c$$

$$= \sqrt{25x^2-4} - 2\cos^{-1}\left(\frac{2}{5x}\right) + c$$

We now have the answer back in terms of x.

This first one needed lot's of explanation since it was the first one. The remaining examples won't need quite as much explanation and so won't take as long to work. However, before we move onto more problems let's first address the issue of definite integrals and how the process differs in these cases.

Example: Evaluate the following integral.

$$\int_{\frac{2}{5}}^{\frac{4}{5}} \frac{\sqrt{25x^2-4}}{x}dx$$

Solution: The limits here won't change the substitution so that will remain the same.

$$x = \frac{2}{5}\sec\theta$$

Using this substitution the square root still reduces down to,

$$\sqrt{25x^2-4} = 2\left|\tan\theta\right|$$

However, unlike the previous example we can't just drop the absolute value bars. In this case we've got limits on the integral and so we can use the limits as well as the substitution to determine the range of that we're in. Once we've got that we can determine how to drop the absolute value bars.

Here's the limits of θ.

$$x=\frac{2}{5} \quad \Rightarrow \quad \frac{2}{5}=\frac{2}{5}\sec\theta \quad \Rightarrow \quad \theta=0$$

$$x=\frac{4}{5} \quad \Rightarrow \quad \frac{4}{5}=\frac{2}{5}\sec\theta \quad \Rightarrow \quad \theta=\frac{\pi}{3}$$

So, if we are in the range $\frac{2}{5}\le x\le\frac{4}{5}$ then θ is in the range of $0\le\theta\le\frac{\pi}{3}$ and in this range of θ's tangent is positive and so we can just drop the absolute value bars.

Let's do the substitution. Note that the work is identical to the previous example and so most of it is left out. We'll pick up at the final integral and then do the substitution.

$$\int_{\frac{2}{5}}^{\frac{4}{5}}\frac{\sqrt{25x^2-4}}{x}dx=2\int_0^{\frac{\pi}{3}}\sec^2\theta-1d\theta$$

$$=2(\tan\theta-\theta)\Big|_0^{\pi/3}$$

$$=2\sqrt{3}-\frac{2\pi}{3}$$

Note that because of the limits we didn't need to resort to a right triangle to complete the problem.

Let's take a look at a different set of limits for this integral.

Example: Evaluate the following integral.

$$\int_{-\frac{4}{5}}^{-\frac{2}{5}}\frac{\sqrt{25x^2-4}}{x}dx$$

Solution: Again, the substitution and square root are the same as the first two examples.

$$x=\frac{2}{5}\sec\theta \qquad \sqrt{25x^2-4}=2|\tan\theta|$$

Let's next see the limits for this problem.

$$x=-\frac{2}{5} \quad \Rightarrow \quad -\frac{2}{5}=\frac{2}{5}\sec\theta \quad \Rightarrow \quad \theta=\pi$$

$$x=-\frac{4}{5} \quad \Rightarrow \quad -\frac{4}{5}=\frac{2}{5}\sec\theta \quad \Rightarrow \quad \theta=\frac{2\pi}{3}$$

Note that in determining the value of θ we used the smallest positive value. Now for this range of x's we have $\frac{2\pi}{3}\le\theta\le\pi$ and in this range of θ tangent is negative and so in this case we can drop the absolute value bars, but will need to add in a minus sign upon doing so. In other words,

$$\sqrt{25x^2-4}=-2\tan\theta$$

So, the only change this will make in the integration process is to put a minus sign in front of the integral. The integral is then,

$$\int_{\frac{2}{5}}^{\frac{4}{5}}\frac{\sqrt{25x^2-4}}{x}dx=-2\int_0^{\frac{\pi}{3}}\sec^2\theta-1d\theta$$

$$=2\left(\tan\theta-\theta\right)\Big|_{2\pi/3}^{\pi}$$

$$=\frac{2\pi}{3}-2\sqrt{3}$$

In the last two examples we saw that we have to be very careful with definite integrals. We need to make sure that we determine the limits on θ and whether or not this will mean that we can drop the absolute value bars or if we need to add in a minus sign when we drop them. Before moving on to the next example let's get the general form for the substitution that we used in the previous set of examples.

$$\sqrt{b^2x^2-a^2} \quad \Rightarrow \quad x=\frac{a}{b}\sec\theta$$

Let's work a new and different type of example.

Example: Evaluate the following integral.

$$\int\frac{1}{x^4\sqrt{9-x^2}}dx$$

Solution: Now, the square root in this problem looks to be (almost) the same as the previous ones so let's try the same type of substitution and see if it will work here as well.

$$x = 3\sec\theta$$

Using this substitution the square root becomes,

$$\sqrt{9-x^2} = \sqrt{9-9\sec^2\theta} = 3\sqrt{1-\sec^2\theta} = 3\sqrt{-\tan^2\theta}$$

So using this substitution we will end up with a negative quantity (the tangent squared is always positive of course) under the square root and this will be trouble. Using this substitution will give complex values and we don't want that. So, using secant for the substitution won't work.

However, the following substitution (and differential) will work.

$$x = 3\sin\theta$$
$$dx = 3\cos\theta d\theta$$

With this substitution the square root is,

$$\sqrt{9-x^2} = 3\sqrt{1-\sec^2\theta} = 3\sqrt{\cos^2\theta} = 3|\cos\theta| = 3\cos\theta$$

We were able to drop the absolute value bars because we are doing an indefinite integral and so we'll assume that everything is positive. The integral is now,

$$\int \frac{1}{x^4\sqrt{9-x^2}}dx = \int \frac{1}{81\sin^4\theta(3\cos\theta)}3\cos\theta d\theta$$
$$= \frac{1}{81}\int \frac{1}{\sin^4\theta}d\theta$$
$$= \frac{1}{81}\int \csc^4\theta d\theta$$

In the previous section we saw how to deal with integrals in which the exponent on the secant was even and since cosecants behave an awful lot like secants we should be able to do something similar with this.

Here is the integral.

$$\int \frac{1}{x^4\sqrt{9-x^2}}dx = \frac{1}{81}\int \csc^2\theta\csc^2\theta d\theta$$

$$= \frac{1}{81}\int(\cot^2\theta + 1)\csc^2\theta\, d\theta \qquad u = \cot\theta$$

$$= \frac{1}{81}\int u^2 + 1 du$$

$$= -\frac{1}{81}\left(\frac{1}{3}\cot^3\theta + \cot\theta\right) + c$$

Now we need to go back to x's using a right triangle. Here is the right triangle for this problem and trig functions for this problem.

$$\sin\theta = \frac{x}{3} \qquad \cos = \frac{\sqrt{9-x^2}}{x}\theta$$

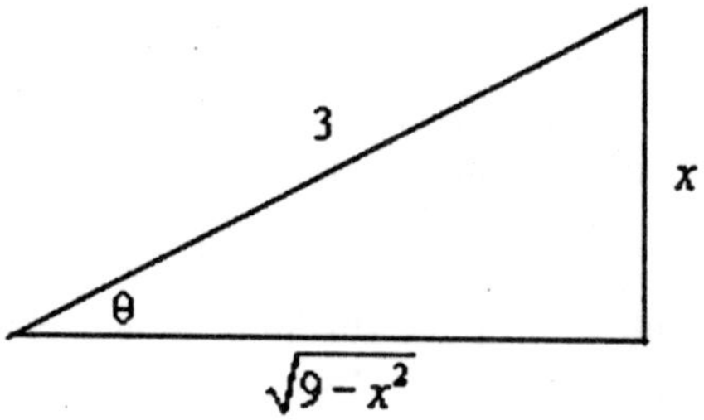

The integral is then,

$$\int \frac{1}{x^4\sqrt{9-x^2}}dx = -\frac{1}{81}\left(\frac{1}{3}\left(\frac{\sqrt{9-x^2}}{x}\right)^3 + \frac{\sqrt{9-x^2}}{x}\right) + c$$

$$= -\frac{(9-x^2)^{\frac{3}{2}}}{243x^3} - \frac{\sqrt{9-x^2}}{81x} + c$$

Here's the general form for this type of square root.

$$\sqrt{a^2 - b^2x^2} \qquad \Rightarrow \qquad x = \frac{a}{b}\sin\theta$$

There is one final case that we need to look at. The next integral will also contain something that we need to make sure we can deal with.

Example: Evaluate the following integral.

$$\int_0^{\frac{1}{6}} \frac{x^5}{\left(36x^2+1\right)^{\frac{3}{2}}} dx$$

Solution: First, notice that there really is a square root in this problem even though it isn't explicitly written out. To see the root let's rewrite things a little.

$$\left(36x^2+1\right)^{\frac{3}{2}} = \left(\left(36x^2+1\right)^{\frac{1}{2}}\right)^3 = \left(\sqrt{36x^2+1}\right)^3$$

This square root is not in the form we saw in the previous examples. Here we will use the substitution for this root.

$$x = \frac{1}{2}\tan\theta \qquad dx = \frac{1}{6}\sec^2\theta d\theta$$

With this substitution the denominator becomes,

$$\left(36x^2+1\right)^3 = \left(\sqrt{\tan^2\theta+1}\right)^3 = \left(\sqrt{\sec^2\theta}\right)^3 = \left|\sec\theta\right|^3$$

Now, because we have limits we'll need to convert them to θ so we can determine how to drop the absolute value bars.

$$x=0 \quad \Rightarrow \quad 0 = \frac{1}{6}\tan\theta \quad \Rightarrow \quad \theta = 0$$

$$x=\frac{1}{6} \quad \Rightarrow \quad \frac{1}{6} = \frac{1}{6}\tan\theta \quad \Rightarrow \quad \theta = \frac{\pi}{4}$$

In this range of θ secant is positive and so we can drop the absolute value bars.

Here is the integral,

$$\int_0^{\frac{1}{6}} \frac{x^5}{(36x^2+1)^{\frac{3}{2}}} dx = \int_0^{\frac{\pi}{4}} \frac{\frac{1}{7776}\tan^5\theta}{\sec^2\theta}\left(\frac{1}{6}\sec^2\theta\right) d\theta$$

$$= \frac{1}{46656}\int_0^{\frac{\pi}{4}} \frac{\tan^5\theta}{\sec\theta} d\theta$$

There are several ways to proceed from this point. Normally

with an odd exponent on the tangent we would strip one of them out and convert to secants.

However, that would require that we also have a secant in the numerator which we don't have. Therefore, it seems like the best way to do this one would be to convert the integrand to sines and cosines.

$$\int_0^{\frac{1}{6}} \frac{x^5}{(36x^2+1)^{\frac{3}{2}}}\,dx = \frac{1}{46656}\int_0^{\frac{\pi}{4}} \frac{\sin^5\theta}{\cos^4\theta}\,d\theta$$

$$= \frac{1}{46656}\int_0^{\frac{\pi}{4}} \frac{(1-\cos^2\theta)^2}{\cos^4\theta}\sin\theta\,d\theta$$

We can now use the substitution $u = cos\ \theta$ and we might as well convert the limits as well.

$$\theta = 0 \qquad u = \cos 0 = 1$$

$$\theta = \frac{\pi}{4} \qquad u = \cos\frac{\pi}{4} = \frac{\sqrt{2}}{2}$$

The integral is then,

$$\int_0^{\frac{1}{6}} \frac{x^5}{(36x^2+1)^{\frac{3}{2}}}\,dx = -\frac{1}{46656}\int_1^{\frac{\sqrt{2}}{2}} u^{-4} - 2u^{-2} + 1\,du$$

$$= \frac{1}{46656}\left(-\frac{1}{3u} + \frac{2}{u} + u\right)\Bigg|_1^{\frac{\sqrt{2}}{2}}$$

$$= \frac{1}{17496} - \frac{11\sqrt{2}}{279936}$$

The general form for this final type of square root is

$$\sqrt{a^2+b^2x^2} \Rightarrow x = \frac{a}{b}\tan\theta$$

We have a couple of final examples to work in this section. Not all trig substitutions will just jump right out at us.

Sometimes we need to do a little work on the integrand first to get it into the correct form and that is the point of the remaining examples.

Example: Evaluate the following integral.

$$\int \frac{x}{\sqrt{2x^2-4x-7}}dx \int \frac{x}{\sqrt{2x^2-4x-7}}dx$$

Solution: In this case the quantity under the root doesn't obviously fit into any of the cases we looked at above and in fact isn't in the any of the forms we saw in the previous examples. Note however that if we complete the square on the quadratic we can make it look somewhat like the above integrals.

Remember that completing the square requires a coefficient of one in front of the x^2.

Once we have that we take half the coefficient of the x, square it, and then add and subtract it to the quantity. Here is the completing the square for this problem.

$$d\left(x^2-2x-\frac{7}{2}\right)=2\left(x^2-2x+1-1-\frac{7}{2}\right)$$

$$=2\left((x-1)^2-\frac{9}{2}\right)=2(x-1)^2-9$$

So, the root becomes,

$$\sqrt{2x^2-4x-7}=\sqrt{2(x-1)^2-9}\sqrt{2x^2-4x-7}$$

$$=\sqrt{2(x-1)^2-9}$$

This looks like a secant substitution except we don't just have an x that is squared. That is okay, it will work the same way.

$$\sqrt{2x^2-4x-7}=\sqrt{2(x-1)^2-9}=\sqrt{2\sec^2\theta-9}$$

$$=3\sqrt{\tan^2\theta}=3\,|\tan\theta|=3\tan\theta$$

$$x-1=\frac{3}{\sqrt{2}}\sec\theta \quad x=1+\frac{3}{\sqrt{2}}\sec\theta \quad dx$$

$$=\frac{3}{\sqrt{2}}\sec\theta\tan\theta\, d\,\theta$$

Using this substitution the root reduces to, Note we could drop the absolute value bars since we are doing an indefinite integral. Here is the integral.

$$\int \frac{x}{\sqrt{2x^3 - 4x - 7}}\,dx = \int \frac{1 + \frac{3}{\sqrt{2}}\sec\theta}{3\tan\theta}\left(\frac{3}{\sqrt{2}}\sec\theta\tan\theta\right)$$

$$= \int \frac{1}{\sqrt{2}}\sec\theta + \frac{3}{2}\sec^2\theta\, d\theta$$

$$= \frac{1}{\sqrt{2}}\ln|\sec\theta + \tan\theta| + \frac{3}{2}\tan\theta + c$$

And here is the right triangle for this problem.

$$\sec\theta = \frac{\sqrt{2}(x-1)}{3} \qquad \tan\theta = \frac{\sqrt{2x^3 - 4x - 7}}{3}$$

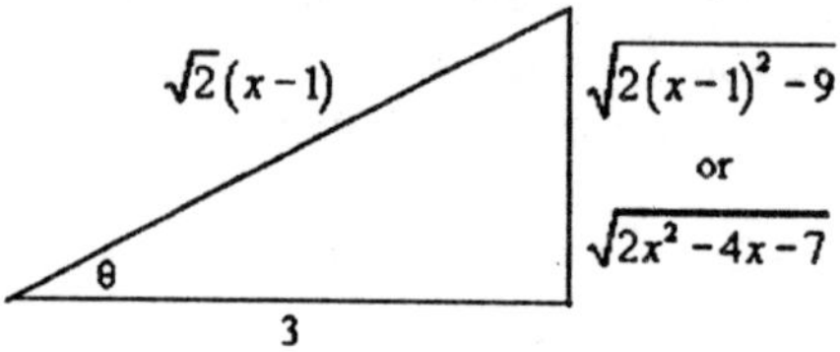

The integral is then,Here is the right triangle for this integral.

$$\int \frac{x}{\sqrt{2x^2 - 4x - 7}}\,dx = \frac{1}{\sqrt{2}}\ln\left|\frac{\sqrt{2}(x-1)}{3} + \frac{\sqrt{2x^2 - 4x - 7}}{3}\right|$$

$$+ \frac{\sqrt{2x^2 - 4x - 7}}{2} + c$$

3

Elementary Integrals

An equivalent integration formula corresponds to everyone of the earlier proved differentiation formulae. Since these elementary integrals are used over and over again as material in the art of integration, we have collected them in the table below. The right-hand column contains elementary functions, the left-hand column the corresponding derivatives.

If we read the table from the left hand side to the right hand side, we find in the right hand column an indefinite integral of the function in the left hand column. We also remind the reader of the fundamental theorems of the differential and integral calculus, and, in particular, of the fact that the definite integral is obtained from the indefinite integral $F(x)$ by the formula

$$\int_a^b f(x)dx = F(x)\Big|_a^b = F(b) - F(a)$$

Finally, for the technique of integration, the readers should have the elementary rules of integration at their finger-tips. In the following sections, we shall attempt to reduce the calculation of integrals of given functions in some way or other to the elementary integrals collected in the table above.

Apart from devices which the beginner certainly could not acquire systematically, but which, on the contrary, occur only to those with long experience, this reduction is based essentially on two useful methods. Each of these methods enables us to transform a given integral in many ways; the objective of such transformations is to reduce the given integral, in one step or in a sequence of steps, to one or more of the above elementary integration formulae.

THE METHOD OF SUBSTITUTION

The first of these useful methods for attacking integration problems is the introduction of a new variable (i.e. the method of substitution or transformation). The corresponding integral formula is just the chain rule of the differential calculus, expressed in the integral form.

The Substitution Formula

We assume that a new variable u is introduced into a function $F(x)$ by means of the equation $x = \phi(u)$, so that $F(x)$ becomes a function of u:

$$F(x) = F(\phi(u)) = G(u)$$

By the chain rule of the differential calculus,

$$\frac{dG}{du} = \frac{dF}{dx}\phi'(u)$$

If we now write

$$F'(x) = f(x)$$

and

$$G'(u) = g(u)$$

or the equivalent expressions

$$F(x) = \int f(x)dx$$

and

$$G(u) = \int g(u)du$$

then, on the one hand, the chain rule takes the form

$$g(u) = \int(x)\phi'(u)$$

and, on the other hand, $G(u) = F(x)$ by definition, that is,

$$\int g(u)du = \int f(x)dx$$

and we obtain the integral formula, equivalent to the chain rule,

$$\int f(\phi(u))\phi'(u)du = \int f(x)dx, \{x = \phi(u)\}$$

This is the basic formula for the substitution of a new variable in an integral. It means that, if we wish to find an indefinite integral of a function of u, which is given in the special form $f(\phi(u))\phi'(u)$, we can instead find the indefinite integral of the function $f(x)$ as a function of x and after integration return to the variable u by setting $x = \phi(u)$. For example, if we apply the formula to the integrand $\phi'(u)/\phi(u)$, we obtain

$$\int\frac{\phi'(u)}{\phi(u)}du = \int\frac{dx}{x} = \log|x| = \log|\phi(u)|$$

or, replacing u by x,

$$\int\frac{\phi'(x)}{\phi(x)}dx = \log|\phi(x)|$$

If we substitute in this important formula particular functions such as $\phi(x) = \log x$ or $\phi(x) = \sin x$, we obtain

$$\int\frac{dx}{x\log x} = \log|\log x|$$

$$\int\cot x dx = \log|\sin x|, \int\tan x dx = -\log|\cos x|$$

A further example is

$$\int\varphi(u)\varphi'(u)du = \int x dx = \frac{1}{2}x^2 = \frac{1}{2}[\varphi(u)]^2$$

where $f(x) = x$. This yields for $\phi(u) = \log u$

$$\int\frac{\log u}{u}du = \frac{1}{2}(\log u)^2$$

These and the following formulae are verified by showing that differentiation of the result returns the integrand. Moreover, the formulae are, of course, only asserted as true in as far as the expressions occurring in them have a meaning. We finally consider the example

$$\int\sin^n u\cos u du$$

Here $x = \sin u = \phi(u)$, whence

$$\int\sin^n u\cos u du = \int x^n dx = \frac{x^{n+1}}{n+1} = \frac{\sin^{n+1} u}{n+1}$$

However, in many cases, we shall use the above formula in the reverse direction, starting with the right hand side - the integral $\int f(x)dx$.

We now have to evaluate or simplify a prescribed indefinite integral $F(x) = \int f(x)dx$ by introduction of the new integration variable u by means of the transformation formula $x = \phi(u)$, then working out the indefinite integral

$$G(u) = \int f(\phi(u))\phi'(u)du$$

and finally replacing the variable u in this integral by x. In order to carry out this last step, we must be certain that a definite value u actually corresponds to the value x, i.e., that the function $x = \phi(u)$ has an inverse. Accordingly, we now make the following assumption, in which we regard x as the primary variable:

In the interval under consideration, $u = \psi(x)$ is a monotonic, differentiable function, the derivative $f'(x)$ of which does not vanish anywhere in the interval. We denote the inverse function, which under these conditions is definite and single-valued, by $x = \phi(u)$; its derivative is then given by $\phi'(u) == 1/\psi'(x)$. As the basic formula for the substitution of a new variable u in an integral, we obtain

$$\int f(x)dx = \int f(\phi(u))\phi'(u)du \quad (u = \psi(x))$$

The indefinite integral $\int f(x)dx$ can be obtained by calculation of the indefinite integral $\int f(\phi(u))\phi'(u)du$ and finally introducing x instead of u for the independent variable by means of the equation $u = \psi(x)$.

Hence it is not sufficient merely to express the old variable x in terms of the new variable u and then to integrate with respect to this new variable; before integrating, we must multiply by this derivative of the original variable x with respect to the new variable u. The corresponding formula for definite integration between two limits is

$$\int_a^b f(x)dx = \int_{\psi(a)}^{\psi(b)} f(\phi(u))\phi'(u)du$$

In the new integral, we have to choose those limits of

integration which are obtained by subjecting the old integration limits to the transformation $x = \phi(u)$, $u = \psi(x)$. In most applications, the integrand $f(x)$ will appear at the outset as a function of a function, say, $f(x) = h(u)$, where $u = \psi(x)$. It is then more convenient to write our integral formula in a slightly different form by identifying the expression $f\{\phi(u)\}$ with the expression $h(h)$. If we make for u the substitution $u = \psi(x)$, $x = \phi(u)$, then our transformation formula is simply

$$\int h\{\psi(x)\}\,dx = \int h(u)\frac{dx}{du}du$$

As a first example, consider the integration of the function $f(x) = \sin 2x$ and introduce $u = \psi(x) = 2x$ and $h(u) = \sin u$. We have

$$\frac{du}{dx} = \psi'(x) = 2$$

If we now substitute $u = 2x$ into the integral as the new variable, then it is not transformed into $\int \sin u\,du$, but into

$$\frac{1}{2}\int \sin u\,du = -\frac{1}{2}\cos u = -\frac{1}{2}\cos 2\pi$$

of course, this may of be verified at once by differentiating the right hand side.

If we integrate for x between the limits 0 and $\pi/4$, the corresponding limits for u are 0 and $\pi/2$, and we obtain

$$\int_0^{\pi/4} \sin 2x\,dx =$$

$$\frac{1}{2}\int_0^{\pi/2} \sin u\,du = -\frac{1}{2}\cos u\Big|_0^{\pi/2} = \frac{1}{2}$$

Another simple example is the integral $\int_1^4 \frac{dx}{\sqrt{x}}$. Here we take $u = \psi(x) = \sqrt{x}$ whence $x = \phi(u) = u$. Since $\phi'(u) = 2u$, we find

$$\int_1^4 \frac{dx}{\sqrt{x}} = 2\int_1^2 \frac{u\,du}{u} = 2\int_1^2 du = 2$$

Proof of the Substitution Formula

Our integration formula can also be explained in another and more direct manner by aiming at the formula for definite integration and basing the proof on the meaning of the definite integral as a limit of a sum. In order to calculate the integral

$$\int_a^b h(\psi(x))dx$$

(for the case $a < b$), we begin with an arbitrary subdivision of the interval $a < x < b$, and then make the sub-division finer and finer. We choose these sub-divisions in the following way.

If the function $u = \psi(x)$ is assumed to be monotonic increasing, there is a (1,1) correspondence between the interval $a < x < b$ on the x-axis and an interval $\alpha < u < \beta$ of the values of u = $\psi(z)$, where $\alpha = \psi(a)$, $\beta = \psi(b)$. We subdivide this interval into n parts of length Δu; there is a corresponding subdivision of the x-interval into subintervals which, in general, are not all of the same length. (The assumption that these subintervals are all equal it by no means essential for the proof!) We denote the points of division of the x-interval by $x_0 = a, x_1, x_2, \ldots, x_n = b$ and the lengths of the corresponding sub-intervals by

$$\Delta x_1, \Delta x_2, \ldots, \Delta x_n$$

The integral under consideration is then the limit of the sum

$$\sum_{\nu=1}^{n} h\{\psi(\xi_\nu)\}\Delta x_\nu$$

where the value ξ_ν is arbitrarily selected from the ν-th subinterval of the x-subdivision. This limit exists (for $\Delta u \to 0$) and is the integral, since on account of the uniform continuity of $x = \phi(u)$ the greatest of the lengths Δx tends to 0 with Δu.

We now write this sum in the form,

$$\sum_{\nu=1}^{n} h(u_\nu)\frac{\Delta x_\nu}{\Delta u}\Delta u \text{ where } u_\nu = \psi(\xi_\nu).$$

By the mean value theorem of the differential calculus,

$$\frac{\Delta x_\nu}{\Delta u} = \phi'(\eta_\nu)$$

where η_ν is a suitably chosen intermediate value of the variable u in the ν-th sub-interval of the u-sub-division and $x = \phi(u)$ denotes the inverse function of $u = \psi(x)$. If we now select the value ξ_v in such a way that ξ and η coincide, i.e., $\xi_v = \phi(\eta_v)$, $\eta_v = \psi(\xi_v)$, then our sum takes the form

$$\sum_{v=1}^{n} h(\eta_v)\phi'(\eta_v)\Delta u$$

If we make here the passage to the limit, we immediately obtain the expression

$$\int_{\alpha}^{\beta} h(u)\frac{dx}{du}du$$

as the limiting value, that is, as the value of the integral under consideration in agreement with the formula given above.

Hence we have proved the theorem: Let $h(u)$ be a continuous function of u in the interval $\alpha < u < \beta$. If the function $u = \psi(x)$ is continuous and monotonic, has a continuous, non-vanishing derivative du/dx in $a < x < b$, and $\psi(a) = \alpha, \psi(b) = \beta$, then

$$\int_{a}^{b} h(\psi(x))dx = \int_{a}^{b} h(u)dx = \int_{\alpha}^{\beta} h(u)\frac{dx}{du}du$$

This formula exhibits the advantage of Leibnitz's notation. In order to carry out the substitution $u=\psi(x)$, we only need write $(dx/du)du$ in place of dx, changing the limits from the original values of x to the corresponding values of u.

Integration Formulae

With the help of the substitution rule, we canevaluate a given integral $\int f(x)dx$ in many cases by reducing it by means of a suitable substitution $x = \phi(u)$ to one of the elementary integrations in our table.

Whether such substitutions exist and how to find them are questions which have no general answer; this is rather a matter in which practice and ingenuity, in contrast to a systematic method, come into their own.

As an example, we shall work out the integral,

$$\int \frac{dx}{\sqrt{(a^2 - x^2)}}$$

using the substitution $x = \phi(u) = au$, $u=\psi(x)=x/a$, $dx=adu$, by which, we obtain,

$$\int \frac{dx}{\sqrt{\left(a^2 - x^2\right)}} = \int \frac{a\,du}{a\sqrt{\left(1 - u^2\right)}} = \text{arc sinu} = \text{arc}\sin\frac{x}{a}, \text{ for}$$

$|x| < |a|$

For the sake of brevity, we have written the symbols *dx* and *du* separately, i.e., $dx = \phi'(u)du$ instead of $dx/du = \phi'(u)$.

The same substitution yields also

$$\int \frac{dx}{a^2 + x^2} =$$

$$\int \frac{a\,du}{a^2\left(1 + u^2\right)} = \frac{1}{a}\arctan u = \frac{1}{a}\arctan\frac{x}{a}$$

$$\int \frac{dx}{\sqrt{\left(a^2 + x^2\right)}} = ar\sinh\frac{x}{a}$$

$$\int \frac{dx}{\sqrt{\left(x^2 - a^2\right)}} = ar\cosh\frac{x}{a}, \text{ for } |x| > |a|$$

$$\int \frac{dx}{a^2 - x^2} = \begin{cases} \dfrac{1}{a} ar\tanh\dfrac{x}{a} \text{ for} |x| < |a| \\ \dfrac{1}{a} ar\coth\dfrac{x}{a} \text{ for} |x| > |a| \end{cases}$$

formulae which occur very frequently and which are easily verified by differentiating the right hand side.

In conclusion, we again emphasize the point: In our substitution process, we have made the assumption that the substitution has a unique inverse $x = \phi(u)$, and indeed that $\psi'(x)$ is nowhere equal to zero in the interval under consideration. If our assumption is not fulfilled, application of the substitution formula may easily lead to wrong conclusions.

If $\psi'(x) = 0$ at isolated points of the interval of integration only, we can avoid these difficulties by subdividing this interval in such a way that $\psi'(x)$ vanishes only at the ends of a sub-interval; we can then apply the substitution to each sub-interval separately.

An application of this method leads at once to the following result, which applies to many special cases: If the derivative $\psi'(x)$ vanishes at a finite number of points, but the function $\psi(x)$remains monotonic, then the substitution formula remains valid.

Further Examples of the Substitution Method

We collect here several examples which the reader should study carefully by way of practice!

By the substitution $u = 1 \pm x$, $du = \pm 2xdx$, we find

$$\int \frac{xdx}{\sqrt{\left(1 \pm x^2\right)}} = \qquad \pm\sqrt{\left(1 \pm x^2\right)}$$

$$\int \frac{xdx}{1 \pm x^2} = \qquad \pm\frac{1}{2}\log\left|1 \pm x^2\right|$$

In these formulae, we must take either the sign + or the sign - in all three places. By the substitution $u = ax + b$, $du = adx$ $(a \neq 0)$, we find,

$$\int \frac{dx}{ax + b} = \qquad \frac{1}{a}\log|ax + b|$$

$$\int (ax + b)^{\alpha} dx =$$

$$\frac{1}{a(\alpha + 1)}(ax + b)^{\alpha + 1} \qquad (\alpha \neq -1)$$

$$\int \sin(ax + b)dx = \qquad -\frac{1}{a}\cos(ax + b)$$

in a similar manner, using the substitution $u = \cos x$, $du = -\sin x\, dx$, we obtain

$$\int \tan xdx = \qquad -\log|\cos x|$$

and by means of the substitution $u = \sin x$, $du = \cos x\, dx$

$$\int \cot xdx = \log|\sin x|$$

Using the analogous substitutions $u = \cosh x$, $du = \sinh x\, dx$ and $u = \sinh x$, $du = \cosh xdx$, we obtain

$$\int \tanh xdx = \qquad \log|\cosh x|$$

$$\log|\sinh x| = \qquad \int \coth xdx$$

By the substitution $u = (a/b)\tan x$, $du = (a/b)\sec x\,dx$, we find the two formulae

$$\frac{1}{b^2}\int\frac{1}{a^2/b^2\tan^2 x+1}\cdot\frac{dx}{\cos^2 x}=\frac{1}{ab}\arctan\left(\frac{a}{b}\tan x\right)$$

and
$$S_m(\alpha) = \frac{1}{m+1}\left[\frac{\sin\frac{(m+1)\alpha}{2}}{\sin\frac{\alpha}{2}}\right]^2$$

We evaluate the integral

$$\int\frac{dx}{\sin xa}$$

by writing $\sin x = 2\sin x/2\cos x/2 = 2\tan x/2\cos x/2$ and setting $u = \tan x/2$ so that $du\sec x/2dx$; the integral then becomes

$$\int\frac{dx}{\sin x} = \int\frac{du}{u}=\log\left|\tan\frac{x}{2}\right|$$

If we replace x by $x + \pi/2$, this formula becomes

$$\int\frac{dx}{\cos x} = \log\left|\tan\left(\frac{x}{2}+\frac{\pi}{4}\right)\right|.$$

If we also apply the known trigonometrical formulae $2\cos x = 1 + \cos 2x$ and $2\sin x = 1 - \cos 2x$, the substitution $u = 2x$ yields

$$\int\cos^2 dx = \frac{1}{2}(x+\sin x\cos x)$$

and
$$\int\sin^2 x dx=\frac{1}{2}(x-\sin x\cos x)$$

By the substitution $x = \cos u$, equivalent to $u = \text{arcos}\, x$ or, more generally,

$x = a\cos u$ $(a \succ 0)$, we can reduce

$$\int\sqrt{(1-x^2)}dx \text{ and } \int\sqrt{(a^2-x^2)}dx$$

respectively, to these formulae. We thus obtain

$$\int\sqrt{(a^2-x^2)}dx = -\frac{a^2}{2}\arccos\frac{x}{a}+\frac{x}{2}\sqrt{(a^2-x^2)}$$

Similarly, the substitution $x = a\cosh u$ yields,

$$\int\sqrt{(x^2-a^2)}dx = -\frac{a^2}{2}\operatorname{arcos}h\frac{x}{a}+\frac{x}{2}\sqrt{(x^2-a^2)}.$$

and the substitution $x = a \sinh u$

$$\int\sqrt{(a^2+x^2)}dx = \frac{a^2}{2}\operatorname{ar}\sin h\frac{x}{a}+\frac{x}{2}\sqrt{(a^2+x^2)}$$

The substitution $u = a/x$, $dx = -(a/u)\,du$ leads to

$$\int\frac{dx}{x\sqrt{(x^2-a^2)}} = -\frac{1}{a}\arcsin\frac{a}{x}$$

$$\int\frac{dx}{x\sqrt{(x^2+a^2)}} = -\frac{1}{a}\operatorname{ar}\sinh\frac{a}{x}$$

$$\int\frac{dx}{x\sqrt{(a^2-x^2)}} = -\frac{1}{a}\operatorname{ar}\cosh\frac{a}{x}$$

Finally, we consider the three integrals

$$\int\sin mx\sin nxdx,\ \int\sin mx\cos nxdx,\ \int\cos mx\cos nxdx$$

where m and n are positive integers. By the well-known trigonometrical formulae

$$\sin mx\sin nx = \frac{1}{2}\{\cos(m-n)x-\cos(m+n)x\}$$

$$\sin mx\cos nx = \frac{1}{2}\{\sin(m+n)x+\sin(m-n)x\}$$

$$\cos mx\cos nx = \frac{1}{2}\{\cos(m+n)x+\cos(m-n)x\}$$

we can divide each of these integrals into two parts. If we now use the substitutions $u= (m + n)\,x$ and $u= (m - n)\,x$, respectively, we obtain directly the system of formulae:

$$\int\sin mx\sin nxdx = \begin{cases} -\frac{1}{2}\left\{\frac{\cos(m+n)x}{m+n}+\frac{\cos(m-n)x}{m-n}\right\} \text{if } m\neq n \\ -\frac{1}{2}\left(x-\frac{\sin 2mx}{2m}\right) \text{if } m=n \end{cases}$$

$$\int\sin mx\cos nxdx =$$

$$\begin{cases} -\frac{1}{2}\left\{\frac{\cos(m+n)x}{m+n}+\frac{\cos(m-n)x}{m-n}\right\}\text{if } m\neq n \\ -\frac{1}{2}\left(\frac{\cos 2mx}{2m}\right)\text{if } m=n \end{cases}$$

$$\int \cos mx \cos nx dx =$$

$$\begin{cases} \frac{1}{2}\left\{\frac{\sin(m+n)x}{m+n}+\frac{\sin(m-n)x}{m-n}\right\}\text{if } m\neq n \\ \frac{1}{2}\left(\frac{\cos 2mx}{2m}+x\right)\text{if } m=n \end{cases}$$

In particular, if we now integrate from $-\pi$ to $+\pi$, we obtain from these formulae the extremely important relations

$$\int_{-\pi}^{+\pi} \sin mx \sin nx dx = \begin{cases} 0 \text{ if } m \neq n \\ \pi \text{ if } m = n \end{cases}$$

$$\int_{-\pi}^{+\pi} \sin mx \cos nx dx = 0$$

$$\int_{-\pi}^{+\pi} \cos mx \cos nx dx = \begin{cases} 0 \text{ if } m \neq n \\ \pi \text{ if } m = n \end{cases}$$

These are the orthogonality relations of the trigonometric functions which we shall meet again in Chap ix.

Integration by Parts

The second useful method for dealing with integration problems is given by the differentiation formula for a product:

$$(fg)' = f'g + fg'$$

If we write this formula as an integral formula, we obtain

$$f(x)g(x) =$$
$$\int g(x)f'(x)dx + \int f(x)g'(x)dx$$

or

$$\int f(x)g'(x)dx = f(x)g(x) - \int g(x)f'(x)dx$$

This formula is referred to as the formula for integration by parts. The calculation of one integral is thereby reduced to the calculation of another integral. In fact, if we split up the integrand of an integral $\int w(x)dx$ into a product $\omega(x) = f(x)\,\phi(x)$ and find the indefinite integral

$$g(x) = \int \phi(x)\,dx$$

of the one factor $\phi(x)$, so that $\phi(x) = g'(x)$, then our formula reduces the integral

$$\int \omega(x)\,dx = \int f(x)\phi(x)\,dx = \int f(x)g'(x)\,dx$$

to the integral $\int g(x)f'(x)\,dx$, which in some cases is more readily evaluated than the original form. Since a given function $\omega(x)$, which occurs as an integrand, can be regarded as a product $f(x)\phi(x) = f(x)g'(x)$ in a great many different ways, this formula provides us with a very effective tool for the transformation of integrals. Written as a formula for definite integration, the formula for integration by parts is

$$\int_a^b f(x)g'(x)\,dx = f(x)g(x)\Big|_a^b - \int_a^b g(x)f'(x)\,dx$$

$$= f(b)g(b) - f(a)g(a) - \int_a^b g(x)f'(x)\,dx$$

In fact, in order to obtain the formula for definite integration from the formula for indefinite integration, we only need replace, first of all, the variable appearing on both sides in the formula for the indefinite integral by the value $x = b$, then by the value $x = a$ and write down the difference of these two expressions.

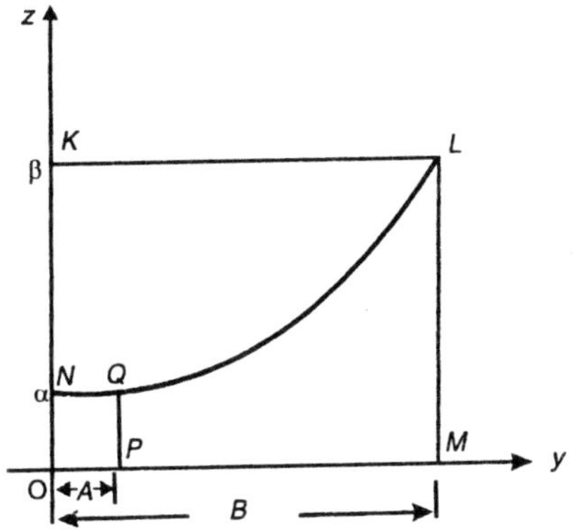

A simple interpretation of this formula, at least with suitable restrictions on the functions involved, can be given. Suppose that $y = f(x)$ and $z = g(x)$ are monotonic, and that $f(x)=A, f(b)=B, g(a)=\alpha$, $g(b)=\beta$; we can then form the inverse of the first function and substitute it into the equation, thus obtaining z as a function of y. Let this function be monotonic increasing. Since $dy = f'(x)dx$ and $dz = g'(x)dx$, the formula for integration by parts can be rewritten

$$\int_A^B zdy + \int_\alpha^\beta ydz = B\beta - A\alpha$$

in agreement with the relation made clear by Figure,

$$\text{area } NQLK + \text{area } PMLQ = \text{area } OMLK - \text{area } OPQN$$

The following example may serve as a first illustration:

$$\int \log x dx = \int \log x.1.dx$$

We write the integrand in this way, in order to indicate that we intend to set $f(x) = \log x$ and $g'(x)=1$, so that we have $f'(x) = 1/x$ and $g(x) = x$. Our formula then becomes

$$\int \log x dx = x\log x - \int \frac{x}{x} dx = x\log x - x$$

This last expression is therefore the integral of the logarithm, as may be verified at once by differentiation.

Recurrence Formulae

In many cases, the integrand is not only a function of the independent variable, but also of an integral index n; on integrating by parts, instead of the value of the integral, we obtain another similar expression in which the index n has a smaller value. We thus arrive, after a number of steps, at an integral, which we can process by means of our table of integrals. Such a procedure is called a recurrence process. The following examples illustrate this method: By repeated integration by parts, we can calculate the trigonometrical integrals

$$\int \cos^n xdx, \int \sin^n xdx, \int \sin^m x\cos^n xdx$$

provided that m and n are integers. In fact, we find that

$$\int \cos^n xdx = \cos^{n-1} x\sin x + (n-1)\int \cos^{n-2} x\sin^2 xdx$$

here we can rewrite the right-hand side in the form,

$$\cos^{n-1} x \sin x + (n-1) \int \cos^{n-2} x dx - (n-1) \int \cos^{n} x dx$$

and thus obtain the recurrence relation,

$$\int \cos^n x dx =$$

$$\frac{1}{n} \cos^{n-1} x \sin x + \frac{n-1}{n} \int \cos^{n-2} x dx$$

This formula enables us to keep on lowering the index in the integrand until we finally arrive at the integral

$$\int \cos x dx = \quad \sin x \text{ or } \int dx = x$$

according to whether n is odd or even. In a similar way, we obtain the analogous recurrence formulae

$$\int \sin^n x dx = -\frac{1}{n} \sin^{n-1} x \cos x + \frac{n-1}{n} \int \sin^{n-2} x dx$$

and

$$\int \sin^m x \cos^n x dx = \frac{\sin^{m+1} x \cos^{n-1} x}{m+n} + \frac{n-1}{m+n} \int \sin^m x cos^{n-2} x dx$$

In particular, these formulae enable us to compute the integrals

$$\int \sin^2 x dx = \quad \frac{1}{2}(x - \sin x \cos)$$

and

$$\int \cos^2 x dx = \quad \frac{1}{2}(x + \sin x \cos x)$$

which we have already obtained by the method of substitution. It need hardly be mentioned that the corresponding integrals for the hyperbolic functions can be calculated in exactly the same manner. Further recurrence relations are obtained by the transformations:

$$\int (\log x)^m dx = x(\log x)^m - m \int (\log x)^{m-1} dx$$

$$\int x^m e^x dx = x^m e^x - m \int x^{m-1} e^x dx$$

$$\int x^m \sin x dx = -x^m \cos x + m \int x^{m-1} \cos x dx$$

$$\int x^m \cos x dx = x^m \sin x - m \int x^{m-1} \sin x dx$$

$$\int x^{\alpha} (\log x)^m dx = \frac{x^{\alpha+1} (\log x)^m}{a+1} - \frac{m}{a+1} \int x^{a} (\log x)^{m-1} dx \quad (a \neq -1)$$

Wallis' Product

The recurrence formula for the integral $\int \sin^n x dx$ leads on an elementary way to a most remarkable expression for the number π as an infinite product. We suppose that $n > 1$ and insertthe limits 0 and $\pi/2$ in the formula

$$\int \sin^n x dx =$$

$$-\frac{1}{n}\sin^{n-1} x \cos x + \frac{n-1}{n}\int \sin^{n-2} x dx$$

obtaining

$$\int_0^{\pi/2} \sin^n x dx = \frac{n-1}{n}\int_0^{\pi/2} \sin^{n-2} x dx \quad \text{for n} > 1$$

Repeated application of the recurrence formula to the right-hand side yields for the cases $n = 2m$ and $n = 2m + 1$

$$\int_0^{\pi/2} \sin^{2m} x dx = \frac{2m-1}{2m}\cdot\frac{2m-3}{2m-2}\cdots\frac{1}{2}\cdot\int_0^{\pi/2} dx$$

$$\int_0^{\pi/2} \sin^{2m+1} x dx = \frac{2m}{2m+1}\cdot\frac{2m-2}{2m-1}\cdots\frac{2}{3}\cdot\int_0^{\pi/2} \sin x dx$$

whence

$$\int_0^{\pi/2} \sin^{2m} x dx = \frac{2m-1}{2m}\cdot\frac{2m-3}{2m-2}\cdots\frac{1}{2}\cdot\frac{\pi}{2}$$

$$\int_0^{\pi/2} \sin^{2m+1} x dx = \frac{2m}{2m+1}\cdot\frac{2m-2}{2m-1}\cdots\frac{2}{3}.$$

Division now yields

$$\frac{\pi}{2} = \frac{2.2}{1.3}\cdot\frac{4.4}{3.5}\cdot\frac{6.6}{5.7}\cdots\frac{2m.2m}{(2m-1)\cdot(2m+1)}\frac{\int_0^{\pi/2} \sin^{2m} x dx}{\int_0^{\pi/2} \sin^{2m+1} x dx}$$

The quotient of the two integrals on the right hand side converges to 1 as m increases, as we recognize from the following considerations. In the interval $0 < x < \pi/2$, we find

$$0 < \sin^{2m+1} x \leqq \sin^{2m} x \leqq \sin^{2m-1} x$$

whence

$$0 < \int_0^{\pi/2} \sin^{2m+1} x dx \leqq \int_0^{\pi/2} \sin^{2m} x dx \leqq \sin^{2m-1} x dx$$

If we here divide each term by $\int_0^{\pi/2} \sin^{2m+1} x dx$ and note that by the first formula, proved above,

$$\frac{\int_0^{\pi/2} \sin^{2m-1} x dx}{\int_0^{\pi/2} \sin^{2m+1} x dx} = \frac{2m+1}{2m} = 1 + \frac{1}{2m}$$

we find that

$$1 \leq \frac{\int_0^{\pi/2} \sin^{2m} x dx}{\int_0^{\pi/2} \sin^{2m+1} x dx} \leq 1 + \frac{1}{2m}$$

which yields the statement above, whence there applies the relation

$$\frac{\pi}{2} = \lim_{m \to \infty} \frac{2}{1}\frac{2}{3}\frac{4}{3}\frac{4}{5}\frac{6}{5}\frac{6}{7} \cdots \frac{2m}{2m-1}\frac{2m}{2m+1}$$

This product formula (due to Wallis), with its simple law of formation, presents a remarkable relation between the number π and the integers.

If we observe that,

$$\lim_{m \to \infty} \frac{2m}{2m+1} = 1$$

we can write and, if we take the square root and then multiply numerator and denominator by 2.4...,... (2m — 2), we find

$$\sqrt{\frac{\pi}{2}} = \lim_{m \to \infty} \frac{2.4...(2m-2)}{3.5...(2m-1)} \sqrt{2m} = \lim_{m \to \infty} \frac{2^2.4^2...(2m-2)^2}{(2m-1)!} \sqrt{2m}$$

$$= \lim_{m \to \infty} \frac{2^2.4^2...(2m)^2}{(2m)!} \frac{\sqrt{2m}}{2m}$$

Hence, we finally obtain

$$\lim_{m \to \infty} \frac{(m!)^2 \, 2^{2m}}{(2m)!\sqrt{m}} = \sqrt{\pi}$$

Integration of Rational Functions

The most important general class of functions, which are integrable in terms of elementary functions, are the rational functions

$$R(x) = \frac{f(x)}{g(x)}$$

where $f(x)$ and $g(x)$ are the polynomials:

$$f(x) = a_m x^m + a_{m-1}x^{m-1} + \ldots + a_0$$
$$g(x) = b_n x^n + a_{n-1}x^{n-1} + \ldots + b_0 \left(b_n \neq 0\right)$$

We recall that every polynomial can be integrated at once and that the integral is itself a polynomial, whence we need only consider those rational functions for which the denominator is not a constant. Moreover, we can always assume that the degree of the numerator is less than the degree (n) of the denominator, because otherwise we can divide the polynomial $f(x)$ by the polynomial $g(x)$ and obtain a remainder of degree less than n; in other words, we can write $f(\mathrm{x})=q(x)g(x)+r(x)$, where $q(x)$ and $r(x)$ are also polynomials and $r(x)$ is of degree less than n. The integration of $f(x)/g(x)$ is then reduced to the integration of the polynomial $q(x)$ and of the proper fraction $r(x)/g(x)$. Moreover, we note that the $f(x)/g(x)$ can be represented as the sum of the functions $a_v x^v/g(x)$, so that we need only consider integrands of the form $x^v/g(x)$.

THE FUNDAMENTAL TYPES

We shall not at once proceed to the integration of the most general rational function of the above type, but instead shall study only those functions in which the denominator $g(x)$ is of a particularly simple type, namely, $g(x) = (\alpha x + \beta)^n$ – a power of a linear $\alpha x+\beta$ $(\alpha \rangle 0)$ – or $g(x) = (ax + 2bx + c)^n$ - a power of a definite * quadratic expression. In the first case, we introduce a new variable $\xi = \alpha x + \beta$. Then $d\xi/dx = \alpha$ and $x = (\xi - \beta)/\alpha$ is also a linear function of ξ. Each numerator $f(x)$ becomes a polynomial $\phi(\xi)$ of the same degree, whence,

$$\int \frac{f(x)}{(\alpha x+\beta)^n} dx = \frac{1}{\alpha} \int \frac{\phi(\xi)}{\xi^n} d\xi$$

A quadratic expression $Q(x) = \alpha x + 2bx + c$ is said to be definite, if it takes for for all real values of x values with one the same sign, i.e., if the equation $Q(x) = 0$ has no real roots. For this, it is necessary and sufficient that $ac - b$ is positive. In the second case, we write,

$$ax^2 + 2bx + c = \frac{1}{a}\left(ax + b^2\right) + \frac{d^2}{a}\left(d^2 = ac - b^2, d > 0\right)$$

observing that, since we have assumed our expression to be definite, $ac - b$ must be positive and $a \nmid 0$. After introducing the new variable

$$\xi = \frac{ax + b}{d}$$

we obtain an integral with the denominator

$$\left[\frac{d^2}{a}\left(1+\xi^2\right)\right]^n$$

Hence, in order to integrate rational functions the denominators of which are powers of a linear expression or of a definite quadratic expression, it is sufficient to be able to integrate the functional types,

$$\frac{1}{x^n}, \frac{x^{2v}}{\left(x^2+1\right)^n}, \frac{x^{2v+1}}{\left(x^2+1\right)^n}$$

In fact, we shall see that, in general, even these types need not be treated, for we can reduce the integration of every rational function to the integration of the very special forms of these three functions, obtained by taking $v = 0$. Hence, we will now consider the integration of the three expressions

$$\frac{1}{x^n}, \frac{1}{\left(x^2+1\right)^n}, \frac{x}{\left(x^2+1\right)^n}$$

Integration of the Fundamental Types

The integration of the first type of function $1/x^n$ yields immediately $\log |x|$, if $n = 1$ and $-1/\{(n-1)x^{n-1}\}$, if $n > 1$, so that in both cases the integral is again an elementary function. Functions of the third type can be integrated immediately after introduction of the new variable $\xi = x + 1$, whence we obtain $2xdx = d\xi$ and

$$\int \frac{x}{\left(x^2+1\right)^n} dx = \frac{1}{2}\int \frac{d\xi}{\xi^n} = \begin{cases} \frac{1}{2}\log\left(x^2+1\right) \text{if } n = 1, \\ -\frac{1}{2(n-1)\left(x^2+1\right)^{n-1}} \text{if } n > 1 \end{cases}$$

Finally, in order to calculate the integral,

$$I_n = \int \frac{dx}{(x^2+1)^{n'}}$$

where n has any value exceeding 1, we use the recurrence method. In fact, if we set

$$\frac{1}{(x^2+1)^n} = \frac{1}{(x^2+1)^{n-1}} - \frac{x^2}{(x^2+1)^n}$$

we can transform the right-hand side by integration by parts, using formulae in 4.3 with

$$f(x) = x,\ g'(x) = \frac{x}{(x^2+1)^n}$$

Then, as we have just discovered above,

$$g(x) = -\frac{1}{2}\frac{1}{(n-1)(x^2+1)^{n-1}}$$

whence we obtain

$$I_n = \int \frac{dx}{(x^2+1)^n} = \frac{x}{2(n-1)(x^2+1)^{n-1}} + \frac{2n-3}{2(n-1)} \int \frac{dx}{(x^2+1)^{n-1}}$$

The calculation of the integral I_n is thus reduced to that of the integral I_{n-1}. If $n - 1 > 1$, we apply the same process to the latter integral and continue the process until we finally arrive at the expression

$$\int \frac{dx}{(x^2+1)} = \text{arc}\tan x$$

We thus see that the integral * I_n can be explicitly expressed in terms of rational functions and the function artan x.

The integral of the function $1/(x-1)^n$ can be calculated in the same way; we reduce it by the corresponding recurrence method to the integral

$$\int \frac{dx}{1-x^2} = \text{ar}\tanh x \ \{or\ \text{ar}\coth x\}$$

Incidentally, we could also have integrated the function $1/(x$

+ 1)^n directly using the substitution x = tanh t; we should then have obtained $dx = \sec t\, dt$ and $1/(1 + x) = \cos t$, so that

$$\int \frac{dx}{\left(x^2+1\right)^n} = \int \cos^{2n-2} t dt$$

and we have already learnt how to evaluate this integral.

Partial Fractions

We are now in a position to integrate the most general rational functions by virtue of the fact that every such function can be represented as the sum of so-called partial fractions, i.e., as the sum of a polynomial and a finite number of rational functions, each one of which has either a power of a linear expression as its denominator and a constant for its numerator or a power of a definite quadratic expression as its denominator and a linear function as its numerator.

If the degree of the numerator $f(x)$ is less than that of the denominator $g(x)$, the polynomial does not occur. We can now integrate each partial fraction, the denominator can be reduced to one of the special form x^n and $(x + 1)^n$, and the fraction is then a combination of the fundamental types integrated.

We shall not give the general proof of the possibility of this resolution into partial fractions. On the contrary, we shall confine ourselves to stating the theorem in an form intelligible for the reader and to showing by examples how the resolution into partial fractions can be carried out in typical cases. In practice, one deals only with comparatively simple functions, since otherwise the computations become far too complicated.

As we know from elementary algebra, every polynomial $g(x)$ can be written in the form,

$$g(x) = a\left(x-a_1\right)^{l_1}\left(x-a_2\right)^{l_2} \ldots \left(x^2+2b_1x+c_1\right)^{r_1}\left(x^2+2b_2x+c_2\right)^{r_2} \ldots$$

Here, the numbers a_1, a_2,... are the real and distinct roots of the equation $g(x)$=0 and the positive integers l_1, l_2, indicate how often they are repeated; the factors $x + 2b_\nu x + c_\nu$ indicate definite quadratic expressions, of which no two are the same, with conjugate complex roots and the positive integers r_1,r_2,... give the numbers of times that these roots are repeated. We assume that the denominator is either given to us in this form or else that we have brought it to this form by calculating the real and imaginary roots. Moreover, let us assume that the numerator $f(x)$ has a lower degree than the denominator. Then the theorem on the resolution into partial fractions is:

For each factor $(x - \alpha)^l$, where α is any one of the real roots and l is the number of times it is repeated, we can determine an expression of the form,

$$\frac{A_1}{(x-a)}+\frac{A_2}{(x-a)^2}+\ldots+\frac{A_l}{(x-a)^{l}}$$

and for each quadratic factor $Q(x) = x + 2bx + c$ in our product, which is raised to the power r, we can determine an expression of the form

$$\frac{B_1+C_1^x}{Q}+\frac{B_2+C_2^x}{Q^2}+\ldots+\frac{B_r+C_r^x}{Q^r}$$

in such a way that the function $f(x)/g(x)$ is the sum of all these expressions. In other words, the quotient $f(x)/g(x)$ can be represented by a sum of fractions, each of which belongs to one or another of the types integrated.

We present here a brief outline of the method by which the possibility of this decomposition into partial fractions is proved. If $g(x) = (x - a)^k h(x)$ and $h(a) \neq 0$, then obviously there vanishes on the right hand side of the equation the numerator for $x = -a$, whence it has the form $h(a)(x - a)^m f_1(x)$, where $f_1(x)$ is also a polynomial, the integer $m \geq 1$ and $f_1(a) \neq 0$. Writing $f(a)/h(a) = \beta$, we obtain

$$\frac{f(x)}{g(x)}=\frac{\beta}{(x-a)^k}=\frac{f_1(x)}{(x-a)^{k-m}h(x)}$$

Continuing the process, we can keep on reducing the degree of the power of $(x - a)$ in the denominator until finally no such factor is left. We repeat for the remaining fraction the process for some other roots of $g(x)$ and do so as many times as g has distinct factors. This is done not only for the real, but also for the complex roots until we arrive at the complete decomposition into partial fractions.

In particular cases, the splitting into partial fractions can be done easily by inspection. If, for example, $g(x) = x - 1$, we see at once that

$$\frac{1}{x^2-1} = \frac{1}{2}\frac{1}{x-1} - \frac{1}{2}\frac{1}{x+1}$$

whence,

$$\int\frac{dx}{x^2-1} = \frac{1}{2}\log\left|\frac{x-1}{x+1}\right|$$

More generally, if $g(x) = (x - \alpha)(x - \beta)$, that is, if $g(x)$ is a non-definite quadratic expression with two real series α and β, we have

$$\frac{1}{(x-\alpha)(x-\beta)} = \frac{1}{\alpha-\beta}\frac{1}{x-\beta}\frac{1}{x-\alpha} - \frac{1}{\alpha-\beta}\frac{1}{x-\beta}$$

whence

$$\int\frac{dx}{(x-\alpha)(x-\beta)} = \frac{1}{\alpha-\beta}\log\left|\frac{x-\alpha}{x-\beta}\right|$$

The Bi-molecular Reaction

A simple example of the application of this easy reduction to partial fractions is given by the so-called bimolecular reaction. Let there be given two reagents the original concentrations of which in mols per until volume are a and b, where we assume that $a < b$; we will assume that there is formed at time t in the unit volume a quantity x (mols) of the reaction product.

Then, according to the law of mass action, in the simplest case - reaction between one molecule of each of the reagents - the rate of increase of the quantity x is given by the equation $dx/dt = k(a - x)\{b - x)$. The problem is to determine the function $x(t)$. Conversely, if we think of the time t as a function of we have integration now yields

$$\frac{dt}{dx} = \frac{1}{k(a-x)(b-x)} = \frac{1}{k(b-a)}\left(\frac{1}{a-x} - \frac{1}{b-x}\right)$$

$$kt = \frac{1}{a-b}\log\frac{a-x}{b-x} + c, \text{for } x < a < b$$

We determine the constant of integration c by the condition that at time $t = 0$ there has not yet formed a product of reaction, so that,

$$\frac{1}{a-b}\log\frac{a}{b}+c = 0$$

Thus, we finally obtain

$$kt = \frac{1}{a-b}\log\frac{1-\frac{x}{a}}{1-\frac{x}{b}}$$

and, by solving for x, the required function

$$x = \frac{ab\left(1-e^{(a-b)kt}\right)}{b-ae^{(a-b)kt}}$$

Further Examples of Resolution into Partial Fractions. The Method of Undetermined Coefficients: If $g(x) = (x-\alpha_1)(x-\alpha_2)\ldots\ldots(x-\alpha_n)$, where $\alpha_i \neq \alpha_k$, $i \neq k$, i.e., if the equation $g(x) = 0$ has only single real roots, the expression in terms of partial fractions has the simple form

$$\frac{1}{g(x)} = \frac{a_1}{x-a_1}+\frac{a_2}{x-a_2}+\ldots+\frac{a_n}{x-a_n}$$

We obtain explicit expressions for the coefficients $a_1, a_2,\ldots$ if we multiply both sides of this equation by $(x - a_1)$, cancel the common factor $(x - a_1)$ in the numerator and denominator on the left hand side and in the first term on the right hand side, and then put $x = a_1$. This yields

$$\alpha_1 = \frac{1}{(\alpha_1-\alpha_2)(\alpha_1-\alpha_3)\ldots(\alpha_1-\alpha_n)}$$

The reader will observe that the denominator on the right hand side is $g'(a_1)$, i.e., the derivative of the function $p(x)$ at the point $x = \alpha_1$. As a typical example of a denominator $g(x)$ with multiple roots, consider the function $1/\{x(x-1)\}$.

The preliminary statement,

$$\frac{1}{x^2(x-1)} = \frac{a}{x-1}+\frac{b}{x}+\frac{c}{x^3}$$

leads us to the required result. If we multiply both sides of this equation by $x(x-1)$, we obtain,

$$1 = (a+b)x^2 - (b-c)x - c$$

which is true for all values of x, from which we have to determine the coefficients a, b, c. This condition cannot hold unless all the coefficients of the polynomial $(a+b)x - (b—c)x - c - 1$ are zero, i.e., we must have

$$a + b = b - c = c + 1 = 0 \text{ or } c = -1, b = -1, a = 1.$$

Thus, we obtain the resolution

$$\frac{1}{x^2(x-1)} = \frac{1}{x-1} - \frac{1}{x} - \frac{1}{x^2}$$

and consequently

$$\int \frac{dx}{x^2(x-1)} = \log|x-1| - \log|x| + \frac{1}{x}$$

We shall now split up the function $1/[x(x + 1)]$ – an example of the case when the zeroes of the denominator are complex - in accordance with the equation

$$\frac{1}{x(x^2+1)} = \frac{a}{x} + \frac{bx+c}{x^2+1}$$

We obtain for the coefficients $a + b = c = a - 1$, so that

$$\frac{1}{x(x^2+1)} = \frac{1}{x} - \frac{x}{x^2+1}$$

and consequently

$$\int \frac{dx}{x(x^2+1)} = \log|x| - \frac{1}{2}\log(x^2+1)$$

We consider as a third example the function $1/(x^4 + 1)$. Even Leibnitz found this to be a troublesome integration. We can represent the denominator as the product of two quadratic factors:

$$x^4 + 1 = (x^2+1)^2 - 2x^2 = \{x^2 + 1 + \sqrt{2x}\}(x^2 + 1 - \sqrt{2x})$$

We now know that the resolution into partial fractions will have the form

$$\frac{1}{x^4+1} = \frac{ax+b}{x^2+\sqrt{2x}+1} + \frac{cx+d}{x^2-\sqrt{2x}+1}$$

In order to determine the coefficients a, b, c, d, we have the equation,

$$(a+c)x^3+\left(b+d-a\sqrt{2}+c\sqrt{2}\right)x^2$$
$$+\left(a+c-b\sqrt{2}+d\sqrt{2}\right)x+(b+d-1)=0$$

which is satisfied by the values,

$$a=\frac{1}{2\sqrt{2}},b=\frac{1}{2},c=-\frac{1}{2\sqrt{2}},d=\frac{1}{2}$$

whence,

$$\frac{1}{x^4+1}=\frac{1}{2\sqrt{2}}\cdot\frac{x+\sqrt{2}}{x^2+\sqrt{2}x+1}-\frac{1}{2\sqrt{2}}\cdot\frac{x-\sqrt{2}}{x^2-\sqrt{2}x+1}$$

and, applying the method, we obtain,

$$\int\frac{dx}{x^4+1}=\frac{1}{4\sqrt{2}}\log\left|x^2+\sqrt{2x}+1\right|-\frac{1}{4\sqrt{2}}\log\left|x^3-\sqrt{2x+1}\right|$$
$$+\frac{1}{2\sqrt{2}}\arctan\left(\sqrt{2x}+1\right)+\frac{1}{2\sqrt{2}}\arctan\left(\sqrt{2x-1}\right)$$

which is readily verified by differentiation.

INTEGRATION OF OTHER CLASSES OF FUNCTIONS

Rational Representation of the Trigonometric and Hyperbolic Functions

The integration of certain other general classes of functions can be reduced to the integratioa of rational functions. We shall better understand this reduction, if we begin by stating certain elementary facts about the trigonometric and hyperbolic functions. If we set t = tan$x$/2, elementary trigonometry yields the simple formulae

$$\sin x = \frac{2t}{1+t^2},\cos x=\frac{1-t^2}{1+t^2}$$

in fact,

$$\frac{1}{1+t^2} = \cos^2\frac{x}{2} \text{ and } \frac{t^2}{1+t^2}=\sin^2\frac{x}{2}$$

whence we obtain the above equations from the elementary formulae,

$$\sin x = 2\cos^2\frac{x}{2}\tan\frac{x}{2}$$

and,

$$\cos x = \cos^2\frac{x}{2} - \sin^2\frac{x}{2}$$

These equations show that sin *x* and cos *x* can be expressed rationally in terms of the quamtity *t*=tan*x*/2. Differentiation now yields,

$$\frac{dt}{dx} = \frac{1}{2\cos^2 x/2} = \frac{1+t^2}{2}$$

whence,

$$\frac{dx}{dt} = \frac{2}{1+t^2}$$

whence the derivative *dx*/*dt* is also a rational expression in *t*.

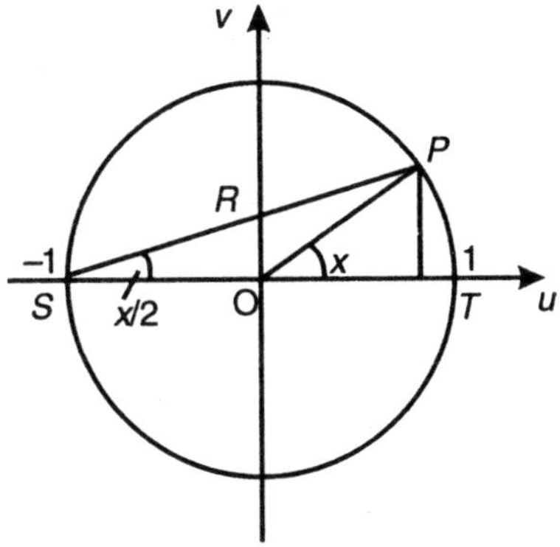

Fig. Parametric Representation of the Trigonometric Functions

Figure shows the geometrical representation and the geometrical meaning of our formulae. It shows the circle $u + v = 1$ in a *uv*-plane.

If *x* denotes the angle *POT*, then $u = \cos x$ and $v = \sin x$. By a theorem in elementaly geometry, the angle *OSP* with its vertex at the point *u*=-1, *v*=0 is equal to *x*/2, and we can deduce from the figure the geometrical meaning of the parameter:

$$t = \tan x/2 = OR.$$

If the point P starts from S and moves once around the circle in the positive direction, i.e., if x moves through the interval from $-\pi$ to $+\pi$, the quantity t will move exactly once throngh the entire range of the values from $-\infty$ to $+\infty$.

We may correspondingly express the hyperbolic functions $\cosh x = (e^x+e^{-x})/2$ and $\sinh x=(e^x-e^{-x})/2$ as rational functions of a third quantity. The most obvious way is to set $e^x = \tau$, so that we have for these hyperbolic functions the rational expressions

$$\cosh x = \frac{1}{2}\left(\tau+\frac{1}{\tau}\right), \sin x = \frac{1}{2}\left(\tau-\frac{1}{\tau}\right)$$

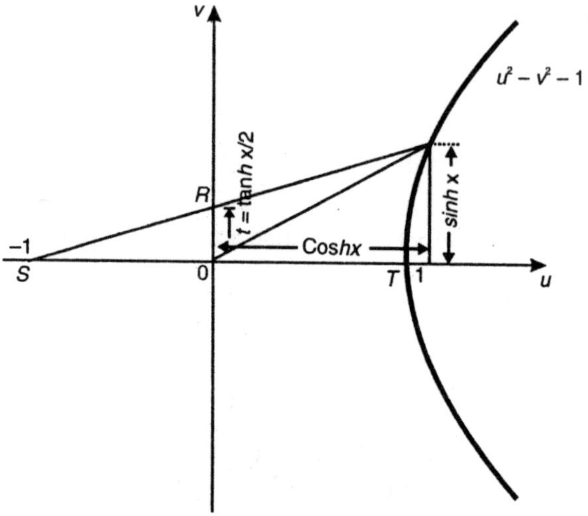

Fig. Parametric Representation of the Hyperbolic Functions

Once again, $dx/dt = 1/\tau$. However, we obtain a closer analogy with the trigonometric functions by introduction of the quantity $t = \tanh x/2$; we then arrive at the formulae

$$\sinh x = \frac{2t}{1-t^2}, \cosh x = \frac{1+t^2}{1-t^2}$$

Differentiation of $t = \tanh x/2$ yields this time the rational expression

$$\frac{dx}{dt} = \frac{2}{1-t^2}$$

for the derivative dx/dt. Once again, the quantity t has a geometrical meaning similar to that which it has in the case of the trigonometric functions, as we see at once from Fig. above

However, while in the case of the trigonometrio functions, t must run through the entire range of values from $-\infty$ to $+\infty$, in order to yield all pairs of values of $\cos x$ and $\sin x$, in the case of the hyperbolic functions, t is limited to the interval $-1 < t < 1$. After these preliminary remarks, we turn to our integration problem.

Integration of R(cos x, sin x)

Let $R(\cos x, \sin x)$ denote an expression which ia rational in the two functions $\sin x$ and $\cos x$, i.e., an expression which is formed rationally from these two functions and given constants such as

$$\frac{2\sin^2 x + \cos x}{3\cos^2 x + \sin x}$$

If we apply the substitution $t = \tan x/2$, the integral

$$\int R(\cos x, \sin x)\, dx$$

becomes the integral

$$\int R\left(\frac{1-t^2}{1+t^2}, \frac{2t}{1+t^2}\right)\frac{2}{1+t^2}\,dt$$

and we have now under the integral sign a rational function of t. Thus, we have obtained theoretically the integral of our expression, since we can now perform the integration by the methods of the preceding section.

Integration of R(cosh x, sinh x)

In the same way, if $R(\cosh x, \sinh x)$ is an expression which is rational in terms of the hyperbolic functions $\cosh x$ and $\sinh x$, we can integrate it by means of the substitution $t = \tanh x/2$. Recalling that

$$\frac{dx}{dt} = \frac{2}{1-t^2}$$

we find

$$\int R(\cosh x, \sinh x)\, dx = \int R\left(\frac{1+t^2}{1-t^2}, \frac{2t}{1-t^2}\right)\frac{2}{1-t^2}\,dt$$

(According to a previous remark, we could also have introduced $\tau = e^x$ as a new variable and expressed cosh x and sinh x in terms of t.) The integration is once again reduced to that of a rational function.

Integration of $R\left\{x, \sqrt{(1-x^2)}\right\}$: The integral $R\left\{x, \sqrt{(1-x^2)}\right\}$ can be reduced by using the substitution

$$x = \cos u, \sqrt{(1-x^2)} = \sin u, dx = -\sin u du$$

the transformation $t = \tan x/2$ leads to the integration of a rational function. By the way, we could have carried out the reduction in a single step by the substitution

$$t = \sqrt{\left(\frac{1-x}{1+x}\right)}; x = \frac{1-t^2}{1+t^2}; \sqrt{(1-x^2)} = \frac{2t}{1+t^2}; \frac{dx}{dt} = \frac{-4t}{(1+t^2)^2}$$

that is, we could have obtained directly a rational function by introduction of a new variable t=tan$u/2$.

Integration of $R\left\{x, \sqrt{(x^2-1)}\right\}$: The integral of this function is transformed by the substitution x = cosh u, when we again introduce

$$t = \sqrt{\left(\frac{x-1}{x+1}\right)} = \tanh\frac{u}{2}$$

Integration of $R\left\{x, \sqrt{(x^2 - 1)}\right\}$

The integral of this function is transformed by the substitution x = sinh u, whence it can be integrated in terms of elementary functions. Instead of the further reduction to an integral of a rational function by the substitution $e^u = \tau$ or tanh $u/2 = t$, we could have reached the integral of a rational function by a single step by either of the substitutions

$$\tau = x + \sqrt{(x^2+1)}, t = \frac{-1+\sqrt{(x^2+1)}}{x}$$

Integration of,

$$R\left\{x, \sqrt{\left(ax^2+2bx+c\right)}\right\}:$$

The integral of this function is rational in terms of x and the square root of an arbitrry polynomial of the second degree in x; it can be reduced immediately to one of the types above.

We write,

$$ax^2+2bx+c = \frac{1}{a}(ax+b)^2+\frac{ac-b^2}{a}$$

If $ac-b>0$, we introduce a new variable $\xi=$ by means of the transformation,

$$\xi=\frac{ax+b}{\sqrt{\left(ac-b^2\right)}}$$

whence the surd takes the form,

$$\sqrt{\left\{\frac{ac-b^2}{a}\left(\xi^2+1\right)\right\}}.$$

The constant a must be positive here, in order that the square root may have real values.

If $ac-b=0$, $a>0$, then we see from,

$$\sqrt{\left(ax^2+2bx+c\right)} = \sqrt{a}\left(x+\frac{b}{a}\right)$$

that the integrand was rational from the start.

Finally, if $ac-b<0$, we set,

$$\xi=\frac{ax+b}{\sqrt{\left(b^2-ac\right)}}$$

and obtain for the surd the expression,

$$\sqrt{\left\{\frac{b^2-ac}{a}\left(\xi^2-1\right)\right\}}.$$

If a is positive, on the other hand, if a is negative, we write the surd in the form

$$\sqrt{\left\{\frac{b^2-ac}{a}\left(\xi^2-1\right)\right\}}.$$

Reduction to Integrals of Rational Functions

We shall briefly mention two other types of functions which can be integrated by reduction to rational functions:

1. Rational expressions involving two different surds of linear expressions

$$R\left\{x, \sqrt{(ax+b)}, \sqrt{(ax+\beta)}\right\}$$

2. Expressions of the form

$$R\left\{x, \sqrt[n]{\left(\frac{ax+b}{ax+\beta}\right)}\right\},$$

where a, b, α β are constants. In (1), we introduce the new variable $\xi = \sqrt{(ax+\beta)}$, so that $\alpha x+\beta = \xi$, whence

$$x = \frac{\xi^2-\beta}{\alpha} \text{ and } \frac{dx}{d\xi} = \frac{2\xi}{\alpha};$$

then

$$\int R\left\{x, \sqrt{(ax+b)}, \sqrt{(ax+\beta)}\right\} dx$$

$$= \int R\left\{\frac{\xi^2-\beta}{\alpha}, \sqrt{\frac{1}{\alpha}\left\{\alpha\xi^2-(\alpha\beta-b\alpha)\right\}}, \xi\right\} \frac{2\xi}{\alpha} d\xi.$$

If we introduce in the second case the new variable

$$\xi = \sqrt[n]{\left(\frac{ax+b}{ax+\beta}\right)}$$

we have

$$\xi^n = \frac{ax+b}{ax+\beta}, x = \frac{-\beta\xi^n+b}{a\xi^n-a}, \frac{dx}{d\xi} = \frac{a\beta-b\alpha}{\left(a\xi^n-\alpha\right)^2} \cdot n\xi^{n-1}$$

and immediately arrive at

$$\int R\left(x, \sqrt[n]{\left(\frac{ax+b}{ax+\beta}\right)}\right) dx = \int R\left(\frac{-\beta\xi^n+b}{\alpha\xi^n-\alpha}, \xi\right) \frac{\alpha\beta-b\alpha}{\left(\alpha\xi^n-\alpha\right)^2} n\xi^{n-1} d\xi$$

which is an integral of a rational function. The preceding discussion is mainly of theoretical interest. In the case of complicated expressions, the actual calculations would be far too involved. It is therefore expedient to employ, whenever it is possible, a special form of the integrand, in order to simplify the work. For example, in order to integrate the expression

$$1/(a\sin x + b\cos x),$$

it is better to use the substitution $t = \tan x$, since $\sin x$ and $\cos x$ can be expressed rationally in terms of $\tan x$ and it is therefore unnecessary to go back to $t = \tan x/2$. The same is true for every expression formed rationally from $\sin x$, $\cos x$ and $\sin x \cos x$*, because $\sin x \cos x = \tan x \cos x$ can, of course, be expanded in terms of $\tan x$. Moreover, for the calculation of many integrals, a trigonometrical form is to be preferred to a rational one, provided that the trigonometrical form can be evaluated by some simple recurrence method. For example, although the integrand in,

$$\int x^n \left\{\sqrt{\left(1-x^2\right)}\right\}^m dx$$

can be reduced to a rational form, it is simpler to write $x = \sin u$ and convert it into $\int \sin^n u \cos^{m+1} u du$ since this can be treated easily by the recurrence method (or by using the addition theorems to reduce the powers of the sine and cosine to sines and cosines of multiple angles). For the evaluation of the integral,

$$\int \frac{dx}{a\cos x + b\sin x}\left(a^2 + b^2 > 0\right)$$

we determine, instead of referring to the general theory, a number A and an angle θ in such a way that

$$a = A\sin\theta, b = A\cos\theta$$

that is, we write

$$A = \sqrt{a^2 + b^2}, \sin\theta = \frac{a}{A}, \cos\theta = \frac{b}{A}$$

The integral then becomes,

$$\frac{1}{A}\int \frac{dx}{\sin\left(x+\theta\right)}$$

and, on introducing the new variable $x+\theta$, yields the value of the integral,

$$\frac{1}{A}\log\left|\tan\frac{x+\theta}{2}\right|$$

FUNCTIONS BY MEANS OF INTEGRALS

Elliptic Integrals

The above examples of types of functions which can be integrated by reduction to rational functions practically exhaust the list of such functions. Attempts to express general integrals such as

$$\int \frac{dx}{\sqrt{\left(a_0 + a_1^x + \ldots + a_n x^n\right)}}, \int \sqrt{\left(a_0 + a_1 x + \ldots + a_n x^n\right)} dx$$

or $\int \frac{e^x}{x} dx$ in terms of elementary functions have always ended up in failure; and finally, in the Nineteenth Century, it has been proved that it is actually impossible to cany out integrations in terms of elementary functions. Hence, if the objective of the integral calculus were integration of functions in terms of elementary functions, we should have come to a definite halt. But such a restricted objective has no intrinsic justiflcation; indeed, it is of a somewhat artificial nature. We know that the integral of every continuous function exists and is itself a continuona function of the upper limit, and this fact has nothing to do with the question whether an integral can be expressed in terms of elementary functions or not.

The distinguishing features of the elementary functions are based on the fact that their properties are easily recognized, that their application to numerical problems is often facilitated by convenient tables or, as in the case of the rational functions, that they are readily calculated with as great a degree of accuracy as we please. Whenever an integral of a function cannot be expressed in terms of functions with which we are already acquainted, there is nothing to stop us from introducing such an integral as a new higher function into analysis, which really means no more than giving it a name.

Whether the introduction of such a new function is convenient or not depends on the its properties, the frequency at which it occurs and the ease with which it can be manipulated in theory and in practice. Hence, in this sense, the process of integration forms a base for the generation of new functions. After all, we are already acquainted with this principle from our dealings with the elementary functions.

The previously unknown integral of $1/x$ as a new function, which we called the logarithm and the properties of which we could easily determine. We could have introduced the trigonometric functions in a similar way, only making use of the rational functions, the process of integration, and the process of inversion. For this purpose, we only need introduce one or the other of the equations

$$\operatorname{arc}\tan x = \int_0^x \frac{dt}{1+t^2} \text{ or } \operatorname{arc}\sin x = \int_0^x \frac{dt}{\sqrt{(1-t^2)}}$$

as the definitions of the function artan x or arsin x, respectively, in order to arrive at the trigonometric functions by inversion. By this process, the definition of these functions becomes separated from geometry, but we are naturally left with the task of also developing their properties independently of geometry. We shall not go here into the development of these ideas. The essential step is to prove the addition theorem for the inverse functions, i,e. for sin and tan.

The first and most important example, which leads us beyond the region of elementary functions, is are the elliptic integrals. These are integrals in which the integrand is formed in a rational way from the variable of integration and the square root of an expression of the third or fourth degree. Among these integrals, the function,

$$u(s) = \int_0^s \frac{dx}{\sqrt{(1-x^2)(1-k^2x^2)}}$$

turns out to be of special importance. Its inverse function $s(u)$ has a correspondingly important role. In particular, for $k = 0$, we obtain $u(s)$ = arsin x and $s(u) = \sin u$, respectively.

The function $s(u)$ has been examined as thoroughly and tabulated as the elementary functions. However, this leads us away from the line of the present discussioa and into the realm of the so-called elliptic functions, which occupy a central position in the theory of functions of a complex variable. We shall merely note here that the name elliptic integral arises from the fact that such integrals enter into the problem of the determination of the length of an arc of an ellipse. Moreover, we may note that integrals which at first sight have quite a different appearance turn out after a simple substitution to be elliptic integrals. For example, the integral

$$\int \frac{dx}{\sqrt{(\cos\alpha - \cos x)}}$$

is transformed by the substitution $u = \cos x/2$ into the integral

$$-k\sqrt{2}\int \frac{du}{\sqrt{(1-u^2)(1-k^2u^2)}}, \qquad k = \frac{1}{\cos\alpha/2}$$

the integral

$$\int \frac{dx}{\sqrt{(\cos 2x)}}$$

by the substitution u = sin x into the integral

$$\int \frac{du}{\sqrt{(1-u^2)(1-2u^2)}}$$

and, finally, the integral

$$\int \frac{dx}{\sqrt{(1-k^2\sin^2 x)}}$$

by the substitution $u = \sin x$ into the integral

$$\int \frac{dx}{\sqrt{(1-u^2)(1-k^2u^2)}}$$

4

Functional Integration

Integration of functions is a summation. If the domain of integration is the square [0, 1] × [0, 1], the integral is computed by breaking the region into small rectangles. Each rectangle serves as the base of a prism whose height is any value of the function within the rectangle. The integral is the sum of the volumes (base × height) of all the prisms. If the rectangles are small enough and the function smooth, the process converges. A functional is a function that associates a function to a number. This is in distinction to common functions that associate numbers to numbers. Examples of functionals include the functional that is one for any function, or the functional that returns the integral of the function over a domain.

There are a number of different ways to approach the Integral Calculus. We will start with the least useful approach, which has the merit of being very simple to understand. We will then look at a more complicated approach which is actually much more useful. Our first approach is going to be to say that Integration is just the opposite (or inverse operation) to Differentiation. If $f(x)$ differentiates to give $F(x)$ then, by definition, $F(x)$ integrates to give $f(x)$. We have been calling $F(x)$ the derivative of $f(x)$ and we will now call $f(x)$ an *Indefinite Integral* of $F(x)$. We use the notation

$$f(x) = \int F(x)\, dx$$

(This rather complicated notation derives from the other approach to integration that we will be considering. Note that both the $\int$ and the dx are part of the notation. If you like, the $\int$ says that we are dealing with an integral and the dx tells us the name of the variable under consideration.) As a simple example, $\sin(x)$ differentiates to give $\cos(x)$, so $\cos(x)$ integrates to give $\sin(x)$. Or

$$\int \cos(x)\, dx = \sin(x)$$

Now we notice a slight problem. $\sin(x)$ differentiates to give $\cos(x)$, but so do $\sin(x) + 1$ and $\sin(x) - 34.23$. In that case which of these is *the* integral of $\cos(x)$? We seem to have an ambiguity. This is true, and it is an ambiguity that will not go away.

How much ambiguity do we have to deal with? Well, suppose that $f_1(x)$ and $f_2(x)$ are both indefinite integrals of $F(x)$. That simply means that both $f_1(x)$ and $f_2(x)$ differentiate to give $F(x)$. But if $f_1'(x) = f_2'(x)$ then, by the rules of differentiation the derivative of $f_1(x) - f_2(x)$ must be zero. So $f_1(x) - f_2(x)$ must be a constant. So the *only* ambiguity is that we can add an arbitrary constant to our integral.

Many authors make this explicit by writing something like

$$\int \cos(x)\, dx = \sin(x) + C$$

where C stands for the arbitrary constant that we can add in. You may have been brought up to do this all the time. You are welcome to do it if that is what you are used to. I will tend to leave it out—simply because I know that it is always there (though there will come a time later in the course when I will have to be more careful).

Let me now gather together some terminology. Consider the formula

$$\int \cos(x)\, dx = \sin(x) + C$$

The function $\cos(x)$ is called the Integrand, the variable x is called the variable of integration, the constant C is called the Constant of Integration and $\sin(x)$ is called the Indefinite Integral or Anti-Derivative.

CALCULATION OF INTEGRALS

We know what integrals are the next problem is to find out

how to do them. This turns out to be a heavy problem. For differentiation we had a tidy set of rules that allowed us to work out the derivative of just about any function that we cared to write down—the procedure is basically mechanical and can be done quite well by computers.

There is nothing like this for integration, in general. Even (apparently) simple integrals can be very difficult or downright impossible to work out. More of this later. Integration is more of a skill than a routine.

One approach that can work sometimes in simple cases is just 'spotting the answer'. If you can spot a function which differentiates to give your function then you have found an integral. Look at the following simple examples.

$$\frac{d}{dx}\left(\frac{1}{n+1}x^{n+1}\right) = x^n \quad \text{so} \quad \int x^n dx = \frac{1}{n+1}x^{n+1} \quad (n \neq -1)$$

$$\frac{d}{dx}\sin x = \cos x \quad \text{so} \quad \int \cos x\, dx = \sin x$$

$$\frac{d}{dx}\cos x = -\sin x \quad \text{so} \quad \int \sin x\, dx = -\cos x$$

$$\frac{d}{dx}e^{ax} = ae^{ax} \quad \text{so} \quad \int e^{ax} dx = \frac{1}{a}e^{ax}$$

$$\frac{d}{dx}\ln x = \frac{1}{x} \quad \text{so} \quad \int \frac{dx}{x} = \ln x$$

Let me do something a little more complicated. We have seen that $\int\cos(x)\,dx = \sin(x)$. What then is the valu e of $\int\cos(2x)\,dx$?

We want to know what function differentiates to give $\cos(2x)$. Well, if you differentiate $\sin(2x)$ you get $2\cos(2x)$, which is not quite right because of the factor 2. That is easily adjusted for and we see that the required integral is $\sin(2x)/2$. Note that this simple approach does not work with an integral like $\int\cos(x^2)\,dx$. The obvious function to play with seems to be $\sin(x^2)$ but the derivative of this is $2x\cos(x^2)$ (damn the chain rule) which is nowhere near the required answer (and no, you can't go dividing by x!). Our simple approach has failed. Your work on differentiation will allow you to check that the following short table is correct. These integrals should be known.

$f(x)$	$\int f(x)dx$
$x^a a \neq 1$	$\frac{1}{a+1} x^a + 1$
$\sin x$	$-\cos(x)$
$\cos(x)$	$\sin(x)$
e^x	e^x
$\frac{1}{x}$	$\text{In}(x)$
$\frac{1}{1+x^2}$	$\arctan x$
$\frac{1}{\sqrt{1-x^2}}$	$\arcsin x$

There is a table of integrals at the end of this booklet. There are books in the library that devote hundreds of pages to listing integrals. Most of the 'techniques' for working out integrals are just methods for changing one integral into another in the hope that you will eventually come to an integral that you can 'spot' or find in the tables.

Let me finish this section by raising (quietly) yet another difficulty with the indefinite integral—it really is plagued with problems. Consider the statement

$$\int \frac{dx}{x} = \ln(x)$$

which is in the above table. This looks correct because we have already found that the derivative of $\ln(x)$ is $1/x$. But look a little closer. The integrand $(1/x)$ makes sense for any value of x other than 0.

The integral $(\ln(x))$ makes no sense at all if x 0. What we have in this formula is really an integral for $1/x$ valid for $x > 0$. We have had to limit the 'domain of definition' of the integrand in order to make sense. This problem will have to be coped with later on.

PROPERTIES OF THE INDEFINITE INTEGRAL

These are immediate consequences of the corresponding properties of derivatives. In each equation there is really an arbitrary constant of integration hanging around.

Let $f(x)$ and $g(x)$ be functions and λa constant. Then

$$\int \lambda f(x) dx = \lambda \int f(x) dx$$

$$\int f(x) + g(x) dx = \int f(x) dx + \int g(x) dx$$

provided that the integrals exist; we have not yet shown that there necessarily exists a function $F(x)$ that differentiates to give a specified function $f(x)$.

These rules may allow us to reduce an integral to the point where we can spot the answer.

$$\int 2x + e^x dx = 2 \int x\, dx + \int e^x dx = x^2 + e^x$$

$$\int \sin x + 2\cos x\, dx = \int \sin x\, dx + 2$$

$$\int \cos x\, dx = -\cos x + 2\sin x$$

IMPOSSIBLE INTEGRALS

You will frequently be told that some integrals are 'impossible' they cannot be done. Common examples are such seemingly respectable integrals as $\int e^{x2}$ dx, $\int \sin(x^2)$ dx, $\int \frac{dx}{\sqrt{x^2+1}} dx$. This does *not* mean that, for example, there is no function that differentiates to give e^{x2}; there certainly is. The problem is that you do not know how to write it down. The answer cannot be expressed in terms of 'elementary functions'.

Let me give another example to make this clearer. We have seen that

$$\int \frac{dx}{1+x^2} = \arctan(x)$$

That is simple enough. But suppose I had not done the inverse trig functions. Then you would never have met arctan(x), so I would have to say to you that, though the integral could certainly be done, it could not be done in terms of functions known to you.

The same kind of thing applies to the 'impossible' examples given above. It is a distressing fact that the vast majority of integrals are, in this sense, impossible. When I write exercise sheets on differentiation I can get away with writing functions down at random, knowing full well that they can be differentiated.

When writing exercises on integration I have to tread very carefully indeed and check everything out before I set anything. In some ways it is *more* surprising that derivatives *can* be done. It just happens to be the case that the derivatives of all the elementary functions can be expressed in terms of elementary functions.

There is no logical reason for this—it is pure luck. That's a lie actually, there are good historical reasons why our idea of 'elementary' for functions leads to simple derivatives.

THE DEFINITE INTEGRAL

Suppose that $F(x)$ is an indefinite integral of $f(x)$, i.e. $F'(x) = f(x)$. The Definite Integral

$$\int_a^b f(x)\,dx$$

where a and b are numbers, is defined to be the number $F(b) - F(a)$:

$$\int_a^b f(x)dx = \left[F(x)\right]_a^b = F(b) - F(a)$$

The choice of indefinite integral (choice of constant of integration) does not matter—the constant of integration cancels out. a and b are called the Limits of Integration. a is called the Lower Limit and b is called the Upper Limit.

Note the 'square bracket' notation,

$$\left[f(x)\right]_a^b$$

stands for $f(b) - f(a)$. Some books use,

$$f(x)\Big|_a^b,$$

but this can be confusing in complicated expressions because it does not tell you where the expression starts. This may all seem a strange idea at this stage. The justification for definite integrals will come quite soon.

DIFFERENTIATING AN INTEGRAL

Consider the function,

$$g(x) = \int_a^x f(t)\,dt$$

where a is a constant and we are thinking of the upper limit of the integral as a variable. If $F(x)$ is an indefinite integral for $f(x)$ then

$$g(x) = F(x) - F(a)$$

So, by definition of $F(x)$,

$$g'(x) = F'(x) = f(x)$$

This gives the result

$$\frac{d}{dx}\int_a^x f(t)\,dt = f(x)$$

PROPERTIES OF THE DEFINITE INTEGRAL

If we assume that the function $f(x)$ has an integral then

$$\int_a^b f(x)\,dx = -\int_b^a f(x)\,dx \quad (F(b) - (F(a) = -(F(a) - (F(b)$$

$$\int_a^b f(x)\,dx + \int_a^c f(x)\,dx = -\int_a^c f(x)\,dx \quad ((F(b) - F(a))$$

$$+ (F(c) - (F(b)) = (F(c) - (F(a))$$

Infinite Limits

We have seen what we mean by a definite integral where

both limits of integration are *finite*. In many important applications of integration you will come across cases where one or more of the limits of integration is infinite. There are three possibilities:

$$\int_a^\infty f(x)\,dx \quad \int_{-\infty}^a f(x)\,dx \quad \int_{-\infty}^\infty f(x)\,dx$$

The interpretation of these is that they are to be taken as limiting cases of the corresponding integrals with finite limits. For example,

$$\int_a^\infty f(x)\,\mathrm{d}x = \lim_{b\to\infty} \int_a^b f(x)\,\mathrm{d}x$$

$$\int_{-\infty}^a f(x)\,\mathrm{d}x = \lim_{b\to-\infty} \int_b^a f(x)\,\mathrm{d}x$$

Here is a simple example: consider the integral $\int_1^\infty \frac{dx}{x^2}$
Now, for any finite positive value of b,

$$\int_1^b \frac{dx}{x^2} = \left[\frac{-1}{x}\right]_1^b = 1 - \frac{1}{b}$$

Now we look at the limiting behaviour as $b\to\infty$. In the limit $1/b\to 0$ and so the value of the integral tends to 1. Therefore

$$\int_1^\infty \frac{dx}{x^2} = 1$$

Of course, in most cases the limits will not exist, so the integral will not exist either. For example,

$$\int_0^\infty x\,\mathrm{d}x = \lim_{a\to\infty} \frac{1}{2}a^2$$

This limit does not exist, so the integral does not exist, though you might get away with saying that it is 'infinite'.

More seriously, consider the integral

$$\int_0^\infty \cos x\,\mathrm{d}x = \lim_{a\to\infty} \sin a$$

Here the limit does not exist either, nor does it do anything so straightforward as going off to infinity—it just oscillates to and fro between −1 and 1.

Finally, let me note a small problem. The integral,

$$\int_{-\infty}^{\infty} f(x)$$

d*x* is the limiting case of,

$$\int_a^b f(x)\,dx \text{ as } a \to \text{¥ and } b \to \text{¥.}$$

This need not be the samè thing as,

$$\lim_{a\to\infty} \int_{-a}^{a} f(x)dx.$$

It is best to think of such doubly-infinite limits in two stages: first let the top limit tend to infinity to get $\int_a^{\infty}$ and then let the bottom limit tend to - ¥ to get the answer.

The reason for such care is shown by the integral of sin(x). For any finite value of a we have,

$$\int_{-a}^{a} \sin(x)\,dx = 0 \text{ (check).}$$

If we took the limit $a \to$ ¥ in this we would get

$$\int_{-\infty}^{\infty} \sin(x)\,dx = 0$$

but this is actually nonsense. If, on the other hand, you start by studying $\int_a^{\infty} \sin(x)\,dx$ you will see at once that no meaningful limiting value exists.

5

Rational Integral

In this section we are going to take a look at integrals of rational expressions of polynomials and once again let's start this section out with an integral that we can already do so we can contrast it with the integrals that we'll be doing in this section.

$$\int \frac{2x-1}{x^2-x+6}\,dx = \int \frac{1}{u}\,du \quad \text{using} \quad u = x^2 - x + 6 \text{ and } du = (2x-1)\,dx$$

$$= \ln|x^2 - x + 6| + c$$

So, if the numerator is the derivative of the denominator (or a constant multiple of the derivative of the denominator) doing this kind of integral is fairly simple. However, often the numerator isn't the derivative of the denominator (or a constant multiple). For example, consider the following integral.

$$\int \frac{2x+11}{x^2-x+6}\,dx$$

In this case the numerator is definitely not the derivative of the denominator nor is it a constant multiple of the derivative of the denominator. Therefore, the simple substitution that we used above won't work. However, if we notice that the integrand can be broken up as follows,

$$\frac{3x+11}{x^2 - \quad - 6} = \frac{4}{x-3} - \frac{1}{x+2}$$

then the integral is actually quite simple.

$$\int \frac{3x+11}{x^2-x-6}\,dx = \int \frac{4}{x-3} - \frac{1}{x+2}\,dx$$

$$= 4\ln|x-3| - \ln|x+2| + c$$

This process of taking a rational expression and decomposing it into simpler rational expressions that we can add or subtract to get the original rational expression is called partial fraction decomposition. Many integrals involving rational expressions can be done if we first do partial fractions on the integrand.

So, let's do a quick review of partial fractions. We'll start with a rational expression in the form,

$$f(x) = \frac{P(x)}{Q(x)}$$

where both *P(x)* and *Q(x)* are polynomials and the degree of *P(x)* is smaller than the degree of *Q(x)*. Recall that the degree of a polynomial is the largest exponent in the polynomial. Partial fractions can only be done if the degree of the numerator is strictly less than the degree of the denominator.

That is important to remember. So, once we've determined that partial fractions can be done we factor the denominator as completely as possible. Then for each factor in the denominator we can use the following table to determine the term(s) we pick up in the partial fraction decomposition.

Factor in Term in partial denominator fraction decomposition

$$ax + b \quad \frac{A}{ax+b} \quad (ax + b)^k$$

$$\frac{A_1}{ax+b} + \frac{A_2}{(ax+b)^2} + \cdots + \frac{A_k}{(ax+b)^k},\ k=1,2,3,....$$

$$ax^2 + bx + c \quad \frac{Ax+B}{ax^2+bx+c} \quad (ax^2 + bx + c)^k$$

$$\frac{A_1x+B_1}{ax^2+bx+c} + \frac{A_2x+B_1}{ax^2+bx+c} + \frac{A_2x+B_2}{\left(ax^2bx+c\right)^2}$$

$$+ \cdots + \frac{A_2x+B_k}{\left(ax^2bx+c\right)^k},\ k=1,2,3,....$$

Notice that the first and third cases are really special cases of the second and fourth cases respectively. There are several methods for determining the coefficients for each term and we will go over each of those in the following examples.

Let's start the examples by doing the integral above.

Example: Evaluate the following integral. $\int \frac{3x+11}{x^2-x-6}dx$

Solution: The first step is to factor the denominator as much as possible and get the form of the partial fraction decomposition. Doing this gives,

$$\frac{3x+11}{(x-3)(x+2)} = \frac{A}{x-3} + \frac{B}{x+2}$$

The next step is to actually add the right side back up.

$$\frac{3x+11}{(x-3)(x+2)} = \frac{A(x+2)+B(x-3)}{(x-3)(x+2)}$$

Now, we need to choose A and B so that the numerators of these two are equal for every x. To do this we'll need to set the numerators equal.

$$3x + 11 = A(x + 2) + B(x - 3)$$

Note that in most problems we will go straight from the general form of the decomposition to this step and not bother with actually adding the terms back up. The only point to adding the terms is to get the numerator and we can get that without actually writing down the results of the addition. At this point we have one of two ways to proceed. One way will always work, but is often more work. The other, while it won't always work, is often quicker when it does work.

In this case both will work and so we'll use the quicker way for this example. We'll take a look at the other method in a later example. What we're going to do here is to notice that the numerators must be equal for *any* x that we would choose to use. In particular the numerators must be equal for $x = -2$and $x = 3$. So, let's plug these in and see what we get.

$$x=-2 \qquad 5 = A(0) + B(-5) \quad \Rightarrow \quad B = -1$$

$$x=3 \qquad 20 = A(5) + B(0) \quad \Rightarrow \quad A = 4$$

So, by carefully picking the x's we got the unknown constants to quickly drop out. Note that these are the values we claimed they would be above.

At this point there really isn't a whole lot to do other than the integral.

$$\int \frac{3x+11}{x^2-x-6}\,dx = \int \frac{4}{x-3} - \frac{1}{x+2}\,dx$$
$$= \int \frac{4}{x-3}\,dx - \int \frac{1}{x+2}\,dx$$
$$= 4\ln|x-3| - \ln|x+2| + c$$

Recall that to do this integral we first split it up into two integrals and then used the substitutions,

$$u = x-3 \quad v = x+2$$

on the integrals to get the final answer.

Before moving onto the next example a couple of quick notes are in order here. First, many of the integrals in partial fractions problems come down to the type of integral seen above. Make sure that you can do those integrals. There is also another integral that often shows up in these kinds of problems so we may as well give the formula for it here since we are already on the subject.

$$\int \frac{1}{x^2+a^2}\,dx = \frac{1}{a}\tan^{-1}\left(\frac{x}{a}\right) + c$$

It will be an example or two before we use this so don't forget about it.

Now, let's work some more examples.

Example: Evaluate the following integral.

$$\int \frac{x^2+4}{3x^2+4x^2-4x}\,dx$$

Solution: We won't be putting as much detail into this solution as we did in the previous example. The first thing is to factor the denominator and get the form of the partial fraction decomposition.

$$\frac{x^2+4}{x(x+2)(3x-2)} = \frac{A}{x} + \frac{B}{x+2} + \frac{C}{3x-2}$$

The next step is to set numerators equal. If you need to

actually add the right side together to get the numerator for that side then you should do so, however, it will definitely make the problem quicker if you can do the addition in your head to get,

$$x^2 + 4 = A\,(x + 2)(3x - 2) + Bx(3x - 2) + Cx(x + 2)$$

As with the previous example it looks like we can just pick a few values of x and find the constants so let's do that.

$$x = 0 \qquad 4 = A(2)(-2) \qquad \Rightarrow \quad A = -1$$

$$x = -2 \qquad 8 = B(-2)(-8) \qquad \Rightarrow \quad B = \frac{1}{2}$$

$$x = \frac{2}{3} \qquad \frac{40}{9} = C\left(\frac{2}{3}\right)\left(\frac{8}{3}\right) \qquad \Rightarrow \quad C = \frac{40}{16} = \frac{5}{2}$$

Note that unlike the first example most of the coefficients here are fractions. That is not unusual so don't get excited about it when it happens.

Now, let's do the integral.

$$\int \frac{x^2 + 4}{3x^3 + 4x^2 - 4x}\,dx = \int -\frac{1}{x} + \frac{\frac{1}{2}}{x + 2} + \frac{\frac{1}{2}}{3x + 2}\,dx$$

$$= -\ln|x| + \frac{1}{2}\ln|x + 2| + \frac{5}{6}\ln|3x - 2| + c$$

Again, as noted above, integrals that generate natural logarithms are very common in these problems so make sure you can do them.

Example: Evaluate the following integral.

$$\int \frac{x^2 - 29x + 5}{(x - 4)^2 (x^2 + 3)}\,dx$$

Solution: This time the denominator is already factored so let's just jump right to the partial fraction decomposition.

$$\frac{x^2 - 29x + 5}{(x - 4)^2 (x^2 + 3)} = \frac{A}{x - 4} + \frac{B}{(x - 4)^2} + \frac{Cx + D}{x^2 + 3}$$

Setting numerators gives,

$$x^2 - 29x + 5 = A(x - 4)(x^2 + 3) + B(x^2 + 3) + (Cx + D)(x - 4)^2$$

In this case we aren't going to be able to just pick values of x that will give us all the constants. Therefore, we will need to work this the second (and often longer) way. The first step is to multiply out the right side and collect all the like terms together. Doing this gives,

$$x^2 - 29x + 5 = (A + C)x^3 + (-4A + B - 8C + D)x^2$$
$$(3A + 16C - 8D)x - 12A + 3B + 16D$$

Now we need to choose *A*, *B*, *C*, and *D* so that these two are equal. In other words we will need to set the coefficients of like powers of *x* equal. This will give a system of equations that can be solved.

$$\left.\begin{aligned} x^3 &: \quad A + C = 0 \\ x^2 &: \quad -4A + B - 8C + D = 1 \\ x^1 &: \quad 3A + 16C - 8D = -29 \\ x^0 &: \quad -12A + 3B + 16D = 5 \end{aligned}\right\} \Rightarrow A = 1, B = -5, C = -1, D = 2$$

Note that we used x^0 to represent the constants. Also note that these systems can often be quite large and have a fair amount of work involved in solving them. The best way to deal with these is to use some form of computer aided solving techniques.

Now, let's take a look at the integral.

$$\int \frac{x^2 - 29x + 5}{(x-4)^2(x^2+3)}\,dx = \int \frac{1}{x-4} - \frac{5}{(x-4)^2} + \frac{-x+2}{x^2+3}\,dx$$
$$= \int \frac{1}{x-4} - \frac{5}{(x-4)^2} - \frac{x}{x^2+3} + \frac{2}{x^2+3}\,dx$$
$$= \ln|x-4| + \frac{5}{x-4} - \frac{1}{4}\ln|x^2+3| + \frac{2}{\sqrt{3}}\tan^{-2}\left(\frac{x}{\sqrt{3}}\right) + c$$

In order to take care of the third term we needed to split it up into two separate terms. Once we've done this we can do all the integrals in the problem. The first two use the substitution, $u = x - 4$ the third uses the substitution $v = x^2 + 3$, and the fourth term uses the formula given above for inverse tangents.

Example: Evaluate the following integral.

$$\int \frac{x^2 + 10x^2 + 3x + 36}{(x-1)(x^2+4)}\,dx$$

Solution: Let's first get the general form of the partial fraction decomposition.

$$\frac{x^3 10x+3x+36}{(x-1)(x^2+4)^2}=\frac{A}{x-1}+\frac{Bx+C}{x^2+4}+\frac{Dx+E}{(x^2+4)}$$

Now, set numerators equal, expand the right side and collect like terms.

$$x^3+10x+3x+36 = A(x^2+4)^2+(Bx+C)(x-1)(x^2+4)+(Dx+E)(x-1).$$
$$=(A+B)x^4+(C-B)x^3+(8A+4B-C+D)x^2+$$
$$(-4B+4C-D+E)x+16A-4C-E$$

Setting coefficient equal gives the following system.

$$\left.\begin{aligned} x^4:&\quad A+B=0\\ x^3:&\quad C+D=1\\ x^2:&\quad 8A+4B-C+D=10\\ x^1:&\quad -4B+4C-D+E=3\\ x^0:&\quad -16A-4C-E=36\end{aligned}\right\}\Rightarrow A=2, B=-2, C=-1, D=1, E=0$$

Don't get excited if some of the coefficients end up being zero. It happens on occasion.

Here's the integral.

$$\int\frac{x^3+10x^2+3x+36}{(x-1)(x^2+4)^2}dx=\int\frac{2}{x-1}+\frac{-2x-1}{x^2+4}+\frac{x}{(x^2+4)^2}dx$$
$$=\int\frac{2}{x-1}-\frac{2x}{x^2+4}-\frac{1}{x^2+4}+\frac{x}{(x^2+4)}dx$$
$$=2\ln|x-1|-\ln|x^2+4|-\frac{1}{2}\tan^{-1}\left(\frac{x}{2}\right)-\frac{1}{2}\frac{1}{x^2+4}+c$$

To this point we've only looked at rational expressions where the degree of the numerator was strictly less that the degree of the denominator.

Of course not all rational expressions will fit into this form and so we need to take a look at a couple of examples where this isn't the case.

Example: Evaluate the following integral.

$$\int \frac{x^4 + 5x^2 + 6x^2 - 18}{x^3 - 3x^2} dx$$

Solution: So, in this case the degree of the numerator is 4 and the degree of the denominator is 3. Therefore, partial fractions can't be done on this rational expression.

To fix this up we'll need to do long division on this to get it into a form that we can deal with. Here is the work for that.

$$\begin{array}{r l} & \underline{\quad x-2 \qquad\qquad\qquad\quad} \\ x^3 - 3x^2 & \big) \, x^4 - 5x^3 + 6x^2 - 18 \\ & \underline{-(x^4 - 3x^3) \qquad\qquad\quad} \\ & \quad -2x^3 + 6x^2 - 18 \\ & \quad \underline{-(-2x^3 + 6x^2) \qquad} \\ & \qquad\qquad\qquad -18 \end{array}$$

So, from the long division we see that,

$$\frac{x^4 - 5x^3 + 6x^2 - 18}{x^3 - 3x^2} = x - 2 - \frac{18}{x^3 - 3x^2}$$

and the integral becomes,

$$\int \frac{x^4 - 5x^3 + 6x^2 - 18}{x^3 - 3x^2} dx = \int x - 2 - \frac{18}{x^3 - 3x^2} dx$$

$$= \int x - 2dx - \int \frac{18}{x^3 - 3x^2} dx$$

The first integral we can do easily enough and the second integral is now in a form that allows us to do partial fractions. So, let's get the general form of the partial fractions for the second integrand.

$$\frac{18}{x^2(x-3)} = \frac{A}{x} + \frac{B}{x^2} + \frac{C}{x-3}$$

Setting numerators equal gives us,

$$18 = Ax(x-3) + B(x-3) + Cx^2$$

Now, there is a variation of the method we used in the first couple of examples that will work here. There are a couple of values of x that will allow us to quickly get two of the three constants, but there is no value of x that will just hand us the third. What we'll do in this example is pick x's to get the two constants that we can easily get and then we'll just pick another value of x that will be easy to work with (*i.e.* it won't give large/messy numbers anywhere) and then we'll use the fact that we also know the other two constants to find the third.

$$x = 0 \qquad 18 = B(-3)(-2) \qquad \Rightarrow \qquad B = -6$$

$$x = 3 \qquad 18 = C(9) \qquad \Rightarrow \qquad C = 2$$

$$x = 1 \qquad 18 = A(-2) + B(-2) + C = -2A + 14 \qquad \Rightarrow \qquad A = -2$$

The integral is then,

$$\int \frac{x^4 - 5x^3 + 6x^2 - 18}{x^3 - 3x^2}\,dx = \int x - 2dx - \int -\frac{2}{x} - \frac{6}{x^2} + \frac{2}{x-3}\,dx$$

$$= \frac{1}{2}x^2 - 2x + 2\ln|x| - \frac{6}{x} - 2\ln|x-3| + c$$

In the previous example there were actually two different ways of dealing with the x^2 in the denominator. One is to treat it as a quadratic which would give the following term in the decomposition

$$\frac{Ax + B}{x^2}$$

and the other is to treat it as a linear term in the following way,

$$x^2 = (x-0)^2$$

which gives the following two terms in the decomposition,

$$\frac{A}{x} + \frac{B}{x^2}$$

We used the second way of thinking about it in our example. Notice however that the two will give identical partial fraction decompositions. So, why talk about this? Simple.

This will work for x^2, but what about x^3 or x^4? In these cases we really will need to use the second way of thinking about these kinds of terms.

$$x^3 \Rightarrow \frac{A}{x} + \frac{B}{x^2} + \frac{C}{x^3} \qquad\qquad x^4 \Rightarrow \frac{A}{x} + \frac{B}{x^2} + \frac{C}{x^3} + \frac{D}{x^4}$$

Let's take a look at one more example.

Example: Evaluate the following integral.

$$\int \frac{x^2}{x^2-1}\,dx$$

Solution: In this case the numerator and denominator have the same degree. As with the last example we'll need to do long division to get this into the correct form.

$$\int \frac{x^2}{x^2-1}\,dx = \int 1+\frac{1}{x^2-1}\,dx = \int dx + \int \frac{1}{x^2-1}\,dx$$

So, we'll need to partial fraction the second integral. Here's the decomposition.

$$\frac{1}{(x-1)(x+1)} = \frac{A}{x-1} + \frac{B}{x-1}$$

Setting numerator equal gives,

$$1 = A\,(x+1) + B(x-1)$$

Picking value of x gives us the following coefficients.

$$x=-1 \qquad 1 = B(-2) \qquad \Rightarrow \qquad B = -\frac{1}{2}$$

$$x=1 \qquad 1 = A(2) \qquad \Rightarrow \qquad A = \frac{1}{2}$$

The integral is then,

$$\int \frac{x^2}{x^2-1}\,dx = \int dx + \int dx \frac{\frac{1}{2}}{x-1} - \frac{\frac{1}{2}}{x-1}\,dx$$

$$= x + \frac{1}{2}\ln|x-1| - \frac{1}{2}\ln|x+1| + c$$

INTEGRALS INVOLVING ROOTS

In this section we're going to look at an integration technique that can be useful for *some* integrals with roots in them. Let's look at a couple of examples to see another technique that can be used on occasion to help with these integrals.

Example: Evaluate the following integral. $\int \frac{x+2}{\sqrt[3]{x-3}}\,dx$

Solution: Sometimes when faced with an integral that contains a root we can use the following substitution to simplify the integral into a form that can be easily worked with.

$$u = \sqrt[3]{x-3}$$

So, instead of letting u be the stuff under the radical as we often did in Calculus I we let u be the whole radical.

Now, there will be a little more work here since we will also need to know what x is so we can substitute in for that in the numerator and so we can compute the differential, dx. This is easy enough to get however. Just solve the substitution for x as follows,

$$x = u^3 + 3 \qquad\qquad dx = 3u^2 du$$

Using this substitution the integral is now,

$$\begin{aligned}\int \frac{(u^3+3)+2}{u} 3u^2 du &= \int 3u^4 + 15u\, du \\ &= \frac{3}{5}u^5 + \frac{15}{2}u^2 + c \\ &= \frac{3}{5}(x-3)^{\frac{5}{3}} + \frac{15}{2}(x-3)^{\frac{2}{3}} + c\end{aligned}$$

So, sometimes, when an integral contains the root $\sqrt[n]{g(x)}$ the substitution,

$$u = \sqrt[n]{g(x)}$$

can be used to simplify the integral into a form that we can deal with.

Let's take a look at another example real quick.

Example: Evaluate the following integral.

$$\int \frac{2}{x - 3\sqrt{x+10}} dx$$

Solution: We'll do the same thing we did in the previous example. Here's the substitution and the extra work we'll need to do to get x in terms of u.

$$u = \sqrt{x+10} \qquad\qquad x = u^2 - 10 \qquad\qquad dx = 2u\, du$$

With this substitution the integral is,

$$\int \frac{2}{x-3\sqrt{x+10}}dx = \int \frac{2}{u^2-10-3u}(2u)du = \int \frac{4u}{u^2-3u-10}du$$

This integral can now be done with partial fractions.

$$\frac{4u}{(u-5)(u+2)} = \frac{A}{u-5} + \frac{B}{u+2}$$

Setting numerators equal gives,

$$4u = A(u+2) + B(u-5)$$

Picking value of u gives the coefficients.

$$u=-2 \qquad -8 = B(-7) \qquad B = \frac{8}{7}$$

$$u=5 \qquad 20 = A(7) \qquad A = \frac{20}{7}$$

The integral is then,

$$\int \frac{2}{x-3\sqrt{x+10}}dx = \int \frac{\frac{20}{7}}{u-5} + \frac{\frac{8}{7}}{u-2}du$$

$$= \frac{20}{7}\ln|u-5| + \frac{8}{7}\ln|u+2| + c$$

$$= \frac{20}{7}\ln|\sqrt{x+10}-5| + \frac{8}{7}\ln|\sqrt{x+10}+2| + c$$

INTEGRALS INVOLVING QUADRATICS

To this point we've seen quite a few integrals that involve quadratics. A couple of examples are,

$$\int \frac{x}{x^2 \pm a}dx = \frac{1}{2}\ln|x^2 \pm a| + c \qquad \int \frac{1}{x^2+a^2}dx = \frac{1}{2}\tan^{-1}\left(\frac{x}{a}\right)$$

We also saw that integrals involving, $\sqrt{b^2x^2-a^2}, \sqrt{a^2-b^2x^2}$ and $\sqrt{a^2+b^2x^2}$ could be done with a trig substitution. Notice however that all of these integrals were missing an x term. They all consist of a quadratic term and a constant. Some integrals involving general quadratics are easy enough to do. For instance, the following integral can be done with a quick substitution.

$$\int \frac{2x+3}{4x^3+12x}dx = \frac{1}{4}\int \frac{1}{u}du \ (u = 4x^2+12x-1 \ \ du = 4(2x+3)dx)$$

$$= \frac{1}{4}\ln|4x^2+12x-1|+c$$

Some integrals with quadratics can be done with partial fractions. For instance,

$$\int \frac{10x-6}{3x^2+16x+5}dx = \int \frac{4}{x+5} - \frac{2}{3x+1}dx = 4\ln|x+5| - \frac{2}{3}\ln|3x+1|+c$$

Unfortunately, these methods won't work on a lot of integrals. A simple substitution will only work if the numerator is a constant multiple of the derivative of the denominator and partial fractions will only work if the denominator can be factored.

This section is how to deal with integrals involving quadratics when the techniques that we've looked at to this point simply won't work. Back in the Trig Substitution section we saw how to deal with square roots that had a general quadratic in them. Let's take a quick look at another one like that since the idea involved in doing that kind integral is exactly what we are going to need for the other integrals in this section.

Example: Evaluate the following integral.

$$\int \sqrt{x^2+4x+5}\,dx$$

Solution: Recall from the Trig Substitution section that in order to do a trig substitution here we first needed to complete the square on the quadratic. This gives,

$$x^2+4x+5 = x^2+4x+4-4+5 = (x+2)^2+1$$

After completing the square the integral becomes,

$$\int \sqrt{x^2+4x+5}dx = \int \sqrt{(x+2)^2+1}\,dx$$

Upon doing this we can identify the trig substitution that we need. Here it is,

$$x+2 = \tan\theta \qquad x = \tan\theta - 2 \qquad dx = \sec^2\theta d\theta$$

$$\sqrt{(x+2)^2+1} = \sqrt{\tan^2\theta+1} = \sqrt{\sec^2\theta} = |\sec\theta| = \sec\theta$$

Recall that since we are doing an indefinite integral we can drop the absolute value bars. Using this substitution the integral becomes,

$$\int \sqrt{x^2 + 4x + 5}\, dx = \int \sec^3 \theta d\theta$$

$$= \frac{1}{2}(\sec\theta \tan\theta + \ln|\sec\theta + \tan\theta|) + c$$

We can finish the integral out with the following right triangle.

$$\tan\theta = \frac{x+2}{1} \qquad \sec\theta = \frac{\sqrt{x^2 4x + 5}}{1} = \sqrt{x^2 + 4x + 5}$$

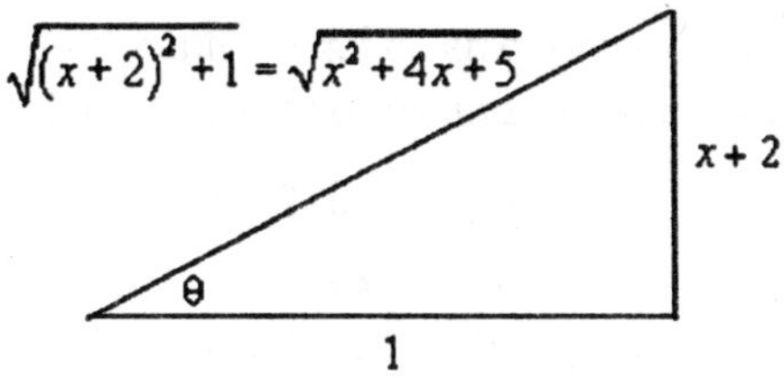

$$\int \sqrt{x^2 + 4x + 5}\, dx = \frac{1}{2}\left((x+2)\sqrt{x^2 + 4x + 5} + \ln|x + 2 + \sqrt{x^2 + 4x + 5}|\right) + c$$

So, by completing the square we were able to take an integral that had a general quadratic in it and convert it into a form that allowed use a known integration technique. Let's do a quick review of completing the square before proceeding. Here is the general completing the square formula that we'll use.

$$x^2 + bx + c = x^2 + bx + \left(\frac{b}{2}\right)^2 - \left(\frac{b}{2}\right)^2 + c = \left(x + \frac{b}{2}\right)^2 + c - \frac{b^2}{4}$$

This will always take a general quadratic and write it in terms of a squared term and a constant term. Recall as well that in order to do this we must have a coefficient of one in front of the x^2. If not we'll need to factor out the coefficient before completing the square. In other words,

$$ax^2 + bx + c = a\left(\underbrace{x^2 + \frac{b}{a}x + \frac{c}{a}}_{\text{complete the square on that}}\right)$$

Now, let's see how completing the square can be used to do integrals that we aren't able to do at this point.

Example: Evaluate the following integral.

$$\int \frac{1}{2x^2 - 3x + 2} dx$$

Solution: Okay, this doesn't factor so partial fractions just won't work on this.

Likewise, since the numerator is just "1" we can't use the substitution. $u = 2x^2 - 3x + 8$. So, let's see what happens if we complete the square on the denominator.

$$\begin{aligned} 2x^2 - 3x + 2 &= 2\left(x^2 - \frac{3}{2}x + 1\right) \\ &= 2\left(x^2 - \frac{3}{2}x + \frac{9}{16} - \frac{9}{16} + 1\right) \\ & 2\left(\left(x - \frac{3}{4}\right)^2 + \frac{7}{16}\right) \end{aligned}$$

With this the integral is,

$$\int \frac{1}{2x^2 - 3x + 2} dx = \frac{1}{2} \int \frac{1}{\left(x - \frac{3}{4}\right)^2 + \frac{7}{16}} dx$$

Now this may not seem like all that great of a change. However, notice that we can now use the following substitution.

$$u = x - \frac{3}{4} \qquad du = dx$$

and the integral is now,

$$\int \frac{1}{2x^2 - 3x + 2} dx = \frac{1}{2} \int \frac{1}{u^2 + \frac{7}{16}} du$$

We can now see that this is an inverse tangent! So, using the formula from above we get,

$$\begin{aligned} \int \frac{1}{2x^2 - 3x + 2} dx &= \frac{1}{2}\left(\frac{4}{\sqrt{7}}\right) \tan^{-1}\left(\frac{4u}{\sqrt{7}}\right) + c \\ &= \frac{2}{\sqrt{7}} \tan^{-1}\left(\frac{4x - 3}{\sqrt{7}}\right) + c \end{aligned}$$

Example: Evaluate the following integral.

$$\int \frac{3x-1}{x^2+10x+28}dx$$

Solution: This example is a little different from the previous one. In this case we do have an x in the numerator however the numerator still isn't a multiple of the derivative of the denominator. So, let's again complete the square on the denominator and see what we get,

$$x^2+10x+28 = x^2+10x+25-25+28 = (x+5)^2+3$$

Upon completing the square the integral becomes,

$$\int \frac{3x-1}{x^2+10x+28}dx = \int \frac{3x-1}{(x+5)^2+3}dx$$

At this point we can use the same type of substitution that we did in the previous example. The only real difference is that we'll need to make sure that we plug the substitution back into the numerator as well.

$$u = x+5 \quad x = u-5 \quad dx = du$$

$$\begin{aligned}\int \frac{3x-1}{x^2+10x+28}dx &= \int \frac{3(u-5)-1}{u^2+3}dx \\ &= \int \frac{3u}{u^2+3} - \frac{16}{u^2+3}du \\ &= \frac{3}{2}\ln|u^2+3| - \frac{16}{\sqrt{3}}\tan^{-1}\left(\frac{u}{\sqrt{3}}\right)+c \\ &= \frac{3}{2}\ln|(x+5)^2+3| - \frac{16}{\sqrt{3}}\tan^{-1}\left(\frac{x+5}{\sqrt{3}}\right)+c\end{aligned}$$

So, in general when dealing with an integral in the form,

$$\int \frac{Ax+B}{dx^2+bx+c}dx$$

Here we are going to assume that the denominator doesn't factor and the numerator isn't a constant multiple of the derivative of the denominator. In these cases we complete the square on the denominator and then do a substitution that will yield an inverse tangent and/or a logarithm depending on the exact form of the numerator.

Let's now take a look at a couple of integrals that are in the same general form except the denominator will also be raised to a power. In other words, let's look at integrals in the form,

$$\int \frac{Ax+B}{(ax^2+bx+c)^n}\,dx$$

Example: Evaluate the following integral.

$$\int \frac{x}{(x^2-6x+11)^3}\,dx$$

Solution: For the most part this integral will work the same as the previous two with one exception that will occur down the road. So, let's start by completing the square on the quadratic in the denominator.

$$x^2-6x+11-x^2--6x+9-9+11=(x-3)^2+2$$

The integral is then,

$$\int \frac{x}{(x^2-6x+11)^3}\,dx = \int \frac{x}{\left[(x-3)^2+2\right]}\,dx$$

Now, we will use the same substitution that we've used to this point in the previous two examples.

$$u=x-3 \qquad x=u+3 \qquad dx=du$$

$$\int \frac{x}{(x^2-6x+11)^3}\,dx = \int \frac{u+3}{(u^2+2)^3}\,du$$

$$= \int \frac{u}{(u^2+2)^3}\,du + \int \frac{3}{(u^2+2)}\,du$$

Now, here is where the differences start cropping up. The first integral can be done with the substitution $v=u^2+2$ and isn't too difficult.

The second integral however, can't be done with the substitution used on the first integral and it isn't an inverse tangent. It turns out that a trig substitution will work nicely on the second integral and it will be the same as we did when we had square roots in the problem.

$$u=\sqrt{2}\tan\theta \qquad du=\sqrt{2}\sec^2\theta\, d\theta$$

With these two substitutions the integrals become,

$$\int \frac{x}{(x^2-6x+11)^3}\,dx = \frac{1}{2}\int \frac{1}{v^3}\,dv + \int \frac{3}{(2\tan^2\theta+2)^2}(\sqrt{2}\sec^2\theta)\,d\theta$$

$$= \frac{1}{4}\frac{1}{v^2} + \int \frac{3\sqrt{2}\sec^2\theta}{8(\tan^2\theta+1)^3}\,d\theta$$

$$= -\frac{1}{4}\frac{1}{(u^2+2)^2} + \frac{3\sqrt{2}}{8}\int \frac{\sec^2\theta}{(\sec^2\theta)^3}\,d\theta$$

$$= \frac{1}{4}\frac{1}{\left((x-3)^2+2\right)} + \frac{3\sqrt{2}}{8}\int \frac{1}{\sec^4\theta}\,d\theta$$

$$= \frac{1}{4}\frac{1}{\left((x-3)^2+2\right)} + \frac{3\sqrt{2}}{8}\int \cos^4\theta\,d\theta$$

Okay, at this point we've got two options for the remaining integral.

We can either use the ideas we learned in the section about integrals involving triginte grals or we could use the following formula.

$$\int \cos^m\theta\,d\theta = \frac{1}{m}\sin\theta\cos^{m-1}\theta + \frac{m-1}{m}\int \cos^{m-2}\theta\,d\theta$$

Let's use this formula to do the integral.

$$\int \cos^4\theta\,d\theta = \frac{1}{4}\sin\theta\cos^3\theta + \frac{3}{4}\int \cos^{m-2}\theta\,d\theta$$

$$= \frac{1}{4}\sin\theta\cos^3\theta + \frac{3}{4}\left(\frac{1}{2}\sin\theta\cos\theta + \frac{1}{2}\int \cos^0\theta\,d\theta\right) \cos{}^\circ\theta = 1!$$

$$= \frac{1}{4}\sin\theta\cos^3\theta + \frac{3}{8}\sin\theta + \frac{3}{8}\sin\theta\cos\theta + \frac{3}{8}\theta$$

Next, let's use the following right triangle to get this back to x's.

$$\tan\theta = \frac{u}{\sqrt{2}} = \frac{x-3}{\sqrt{2}} \quad \sin\theta = \frac{x-3}{\sqrt{(x-3)^2+2}} \quad \cos\theta = \frac{\sqrt{2}}{\sqrt{(x-3)^2+2}}$$

$$\int \cos^4 \theta d\theta = \frac{1}{4}\frac{2\sqrt{2}(x-3)}{\left((x-3)^2+2\right)} + \frac{3}{8}\frac{\sqrt{2}(x-3)}{(x-3)^2+2} + \frac{3}{8}\tan^{-1}\left(\frac{x-3}{\sqrt{2}}\right)$$

$$= \frac{\sqrt{2}}{2}\frac{x-3}{\left((x-3)^2+2\right)^2} + \frac{3\sqrt{2}}{8}\frac{x-3}{(x-3)^2+2} + \frac{3}{8}\tan^{-1}\left(\frac{x-3}{\sqrt{2}}\right)$$

All told then the original integral is,

$$\int \frac{x}{(x^2-6x+11)^3}dx = \frac{1}{4}\frac{1}{\left((x-3)^2+2\right)^2} + \frac{3\sqrt{2}}{8}$$

$$\left(\frac{\sqrt{2}}{2}\frac{x-3}{\left((x-3)^2+2\right)^2} + \frac{3\sqrt{2}}{8}\frac{x-3}{(x-3)^2+2} + \frac{3}{8}\tan^{-1}\left(\frac{x-3}{\sqrt{2}}\right)\right)$$

$$= \frac{1}{8}\frac{3x-11}{\left((x-3)^2+2\right)^2} + \frac{9}{32}\frac{x-3}{(x-3)^2+2} + \frac{9\sqrt{2}}{64}\tan^{-1}\left(\frac{x-3}{\sqrt{2}}\right) + c$$

It's a long and messy answer, but there it is.

Example: Evaluate the following integral.

$$\int \frac{x-3}{(4-2x-x^2)^2}dx$$

Solution: As with the other problems we'll first complete the square on the denominator.

$$4-2x-x^2 = -(x^2+2x-4) = -(x^2+2x+1-1-4)$$
$$= -\left((x+1)^2-5\right) = 5-(x+1)^2$$

The integral is,

$$\int \frac{x-3}{(4-2x-x^2)^2}dx = \int \frac{x-3}{\left[5-(x+1)^2\right]^2}dx$$

Now, let's do the substitution.

$$u = x+1 \qquad x = u-1 \qquad dx = du$$

and the integral is now,

$$\int \frac{x-3}{(4-2x-x^2)^2}dx = \int \frac{u-4}{(5-u^2)^2}du$$
$$= \int \frac{u}{(5-u^2)^2}du - \int \frac{4}{(5-u^2)^2}du$$

In the first integral we'll use the substitution

$$v = 5 - u^2$$

and in the second integral we'll use the following trig substitution

$$u = \sqrt{5}\sin\theta \qquad du = \sqrt{5}\cos\theta d\theta$$

Using these substitutions the integral becomes,

$$\int \frac{x}{(4-2x+x^2)^2}dx = \frac{1}{2}\int \frac{1}{v^2}dv + \int \frac{4}{(5-5\sin^2\theta)^2}(\sqrt{5}\cos\theta)d\theta$$
$$= \frac{1}{2}\frac{1}{v} - \frac{4\sqrt{5}}{25}\int \frac{\cos\theta}{(1-\sin^2\theta)^2}d\theta$$
$$= \frac{1}{2}\frac{1}{v} - \frac{4\sqrt{5}}{25}\int \frac{\cos\theta}{\cos^4\theta}d\theta$$
$$= \frac{1}{2}\frac{1}{v} - \frac{4\sqrt{5}}{25}\int \sin^2\theta d\theta$$
$$= \frac{1}{2}\frac{1}{v} - \frac{2\sqrt{5}}{25}(\sec\theta\tan\theta + \ln|\sec\theta + \tan\theta|) + c$$

We'll need the following right triangle to finish this integral out.

$$\sin\theta = \frac{u}{\sqrt{5}} = \frac{x+1}{\sqrt{5}} \quad \sec\theta = \frac{\sqrt{5}}{\sqrt{5-(x+1)^2}} \quad \tan\theta = \frac{x+1}{\sqrt{5-(x+1)^2}}$$

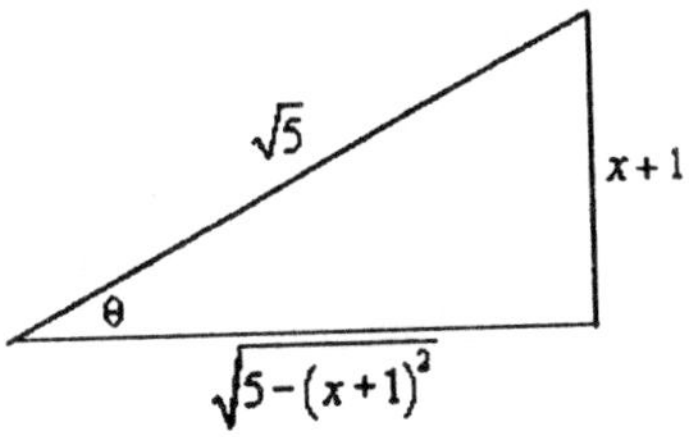

So, going back to x's the integral becomes,

$$\int \frac{x-3}{(4-2x-x^2)^2}dx = \frac{1}{2}\frac{1}{5-u^2}$$

$$-\frac{2\sqrt{5}}{25}\left(\frac{\sqrt{5}(x+1)}{5-(x+1)^2}\ln\left|\frac{\sqrt{5}}{\sqrt{5-(x-1)^2}}+\frac{x+1}{\sqrt{5-(x-1)^2}}\right|\right)+c$$

$$=\frac{1}{10}\frac{4x-1}{5-(x+1)^2}+\frac{2\sqrt{5}}{25}\ln\left|\frac{x+1\sqrt{5}}{\sqrt{5-(x+1)^2}}\right|+c$$

Often the following formula is needed when using the trig substitution that we used in the previous example.

$$\left|\int \sec^m \theta d\theta = \frac{1}{m-1}\tan\theta \sec^{m-2}\theta + \frac{m-2}{m-1}\int \sec^{m-2}\theta d\theta\right|$$

Note that we'll only need the two trig substitutions that we used here. The third trig substitution that we used will not be needed here.

USING INTEGRAL TABLES

This section is entitled Using Integral Tables and we will be using integral tables. However, at some level, this isn't really the point of this section.

To a certain extent the real subject of this section is how to take advantage of known integrals to do integrals that may not look like anything the ones that we do know how to do or are given in a table of integrals.

For the most part we'll be doing this by using substitution to put integrals into a form that we can deal with. However, not all of the integrals will require a substitution. For some integrals all that we need to do is a little rewriting of the integrand to get into a form that we can deal with.

We've already related a new integral to one we could deal with least once. In the last example in the Trig Substitution section we looked at the following integral.

$$\int e^{4x}\sqrt{1+e^{2x}}\,dx$$

At first glance this looks nothing like a trig substitution

problem. However, with the substitution $u = e^x$ we could turn the integral into,

$$\int u^3 \sqrt{1+u^2}\, du$$

which definitely is a trig substitution problem ($u = \tan\theta$). We actually did this process in a single step by using, $e^x = \tan\theta$, but the point is that with a substitution we were able to convert an integral into a form that we could deal with.

So, let's work a couple examples using substitutions and tables.

Example: Evaluate the following integral.

$$\int \frac{\sqrt{7+9x^2}}{x^2}\, dx$$

Solution: So, the first thing we should do is go to the tables and see if there is anything in the tables that is close to this. In the tables in Stewart we find the following integral,

$$\int \frac{\sqrt{a^2+u^2}}{u^2}\, du = \frac{\sqrt{a^2+u^2}}{u} + \ln\left(u + \sqrt{a^2+u^2}\right) + c$$

This is nearly what we've got in our integral. The only real difference is that we've got a coefficient in front of the x^2 and the formula doesn't. This is easily enough dealt with. All we need to do is the following manipulation on the integrand.

$$\int \frac{\sqrt{a^2+u^2}}{u^2}\, dx = \int \frac{\sqrt{9\left(\frac{7}{9}+x^2\right)}}{x^2}\, dx = \int \frac{3\sqrt{\frac{7}{9}+x^2}}{x^2}\, dx = 3\int \frac{\sqrt{\frac{7}{9}+x^2}}{x^2}\, dx$$

So, we can now use the formula with $a = \dfrac{\sqrt{7}}{3}$.

$$\int \frac{\sqrt{7+9x^2}}{x^2}\, dx = 3\left(-\frac{\sqrt{\frac{7}{9}+x^2}}{x} + \ln\left(x + \sqrt{\frac{7}{9}+x^2} \right) \right) + c$$

Example: Evaluate the following integral.

$$\int \frac{\cos x}{\sin x \sqrt{9\sin x - 4}}\, dx$$

Solution: Going through our tables we aren't going to find anything that looks like this in them.

However, notice that with the substitution we can rewrite the integral as,

$$\int \frac{\cos x}{\sin x\sqrt{9\sin x - 4}}\,dx = \int \frac{1}{u\sqrt{9u-4}}$$

and this is in the tables.

$$\int \frac{1}{u\sqrt{a+bu}}\,du = \frac{1}{\sqrt{a}}\ln\left|\frac{\sqrt{a+bu}-\sqrt{a}}{\sqrt{a+bu}+\sqrt{a}}\right| + c \quad \text{if } a > 0$$

$$= \frac{1}{\sqrt{-a}}\tan^{-1}\left(\sqrt{\frac{a+bu}{-a}}\right) + c \quad \text{if } a < 0$$

Notice that this is a formula that will depend upon the value of a. This will happen on occasion. In our case we have $a = -$ and $b = 9$ so we'll use the second formula.

$$\int \frac{\cos x}{\sin x\sqrt{9\sin x - 4}}\,dx = \frac{2}{\sqrt{-(-4)}}\tan^{-1}\left(\sqrt{\frac{9u-4}{-(-4)}}\right) + c$$

$$= \tan^{-1}\left(\sqrt{\frac{9\sin x - 4}{-a}}\right) + c$$

This final example uses a type of formula known as a reduction formula.

Example: Evaluate the following integral.

$$\int \cot^4\left(\frac{x}{2}\right)dx$$

Solution: We'll first need to use the substitution $u = \dfrac{x}{2}$ since none of the formulas in our tables have that in them. Doing this gives,

$$\int \cot^4\left(\frac{x}{2}\right)dx = 2\int \cot^4 u\,du$$

To help us with this integral we'll use the following formula.

$$\int \cot^n u\,du = \frac{-1}{n-1}\cot^{n-1} u - \int \cot^{n-2} u\,du$$

Formulas like this are called reduction formulas. Reduction formulas generally don't explicitly give the integral. Instead they reduce the integral to an easier one. In fact they often reduce the integral to a different version of itself!

For our integral we'll use $n = 4$.

$$\int \cot^4\left(\frac{x}{2}\right)dx = 2\left(-\frac{1}{3}\cot^3 u - \int \cot^{n-2} u\, du\right)$$

At this stage we can either reuse the reduction formula with or use the formula

$$\int \cot^2 u\, du = -\cot u - u + c$$

We'll reuse the reduction formula with $n = 2$ so we can address something that happens on occasion.

$$\int \cot^4\left(\frac{x}{2}\right)dx = 2\left(-\frac{1}{3}\cot^3 u - \left(-\frac{1}{1}\cot u - \int \cot^0 u\, du\right)\right) \quad \cot^0 u = 1!$$

$$= -\frac{2}{3}\cot^3 u + 2\cot u + 2\int du$$

$$= -\frac{2}{3}\cot^3 u + 2\cot u + 2u + c$$

$$= -\frac{2}{3}\cot^3\left(\frac{x}{2}\right) + 2\cot\left(\frac{x}{2}\right) + x + c$$

Don't forget that $a^0 = 1$. Often people forget that and then get stuck on the final integral!

INTEGRATION STRATEGY

We've now seen a fair number of different integration techniques and so we should probably pause at this point and talk a little bit about a strategy to use for determining the correct technique to use when faced with an integral There are a couple of points that need to be made about this strategy.

First, it isn't a hard and fast set of rules for determining the method that should be used. It is really nothing more than a general set of guidelines that will help us to identify techniques that may work. Some integrals can be done in more than one way and so depending on the path you take through the strategy you may end up with a different technique than somebody else who also went through this strategy.

Second, while the strategy is presented as a way to identify the technique that could be used on an integral also keep in mind that, for many integrals, it can also automatically exclude certain techniques as well. When going through the strategy keep two lists in mind. The first list is integration techniques that simply won't work and the second list is techniques that look like they might work. After going through the strategy and the second list has only one entry then that is the technique to use.

If, on the other hand, there are more than one possible technique to use we will then have to decide on which is liable to be the best for us to use. Unfortunately there is no way to teach which technique is the best as that usually depends upon the person and which technique they find to be the easiest.

Third, don't forget that many integrals can be evaluated in multiple ways and so more than one technique may be used on it. This has already been mentioned in each of the previous points, but is important enough to warrant a separate mention. Sometimes one technique will be significantly easier than the others and so don't just stop at the first technique that appears to work. Always identify all possible techniques and then go back and determine which you feel will be the easiest for you to use.

Integrand

This step is very important in the integration process. Many integrals can be taken from impossible or very difficult to very easy with a little simplification or manipulation. Don't forget basic trig and algebraic identities as these can often be used to simplify the integral. We used this idea when we were looking at integrals involving trig functions. For example consider the following integral.

$$\int \cos^2 x \, dx$$

This integral can't be done as is however, simply by recalling the identity,

$$\cos^2 x = \frac{1}{2}(1 + \cos(2x))$$

the integral becomes very easy to do.

Note that this example also shows that simplification does not necessarily mean that we'll write the integrand in a "simpler" form. It only means that we'll write the integrand into a form that we can deal with and this is often longer and/or "messier" than the original integral.

1. See if a "simple" substitution will work. Look to see if a simple substitution can be used instead of the often more complicated methods from Calculus. For example consider both if the following integrals.

   $$\int \frac{x}{x^2-1}dx \qquad \int x\sqrt{x^2-1}dx$$

   The first integral can be done with partial fractions and the second could be done with a trig substitution.

   However, both could also be evaluated using the substitution $u = x^2 - 1$ and the work involved in the substitution would be significantly less than the work involved in either partial fractions or trig substitution.

2. Identify the type of integral. Note that any integral may fall into more than one of these types. Because of this fact it's usually best to go all the way through the list and identify all possible types since one may be easier than the other and it's entirely possible that the easier type is listed lower in the list.

   (a) Is the integrand a rational expression (*i.e* is the integrand a polynomial divided by a poly-nomial)? If so, then partial fractions may work on the integral.

   (b) Is the integrand a polynomial times a trig function, exponential, or logarithm? If so, then integration by parts may work.

   (c) Is the integrand a product of sines and cosines, secant and tangents, or cosecants and cotangents? If so, then the topics from the second section may work.

   Likewise, don't forget that some quotients involving these functions can also be done using these techniques.

   (d) Does the integrand involve $\sqrt{b^2x^2+a^2}$, $\sqrt{b^2x^2-a^2}$, or $\sqrt{a^2 - b^2x^2}$? If so, then a trig substitution might work nicely.

(e) Does the integrand have roots other than those listed above in it? If so, then the substitution $u = \sqrt[n]{g(x)}$ might work.

(f) Does the integrand have a quadratic in it? If so, then completing the square on the quadratic might put it into a form that we can deal with.

3. Can we relate the integral to an integral we already know how to do? In other words, can we use a substitution or manipulation to write the integrand into a form that does fit into the forms we've looked at previously in this chapter.

   A typical example here is the following integral.

$$\int \cos x\sqrt{1+\sin^2 x}\, dx$$

   This integral doesn't obviously fit into any of the forms we looked at in this chapter. However, with the substitution $u = \sin x$ we can reduce the integral to the form,

$$\sqrt{1+u^2}\, du$$

   which is a trig substitution problem.

4. Do we need to use multiple techniques? In this step we need to ask ourselves if it is possible that we'll need to use multiple techniques. The example in the previous part is a good example. Using a substitution didn't allow us to actually do the integral. All it did was put the integral and put it into a form that we could use a different technique on. Don't ever get locked into the idea that an integral will only require one step to completely evaluate it. Many will require more than one step.

5. Try again. If everything that you've tried to this point doesn't work then go back through the process and try again. This time try a technique that that you didn't use the first time around.

As noted above this strategy is not a hard and fast set of rules. It is only intended to guide you through the process of best determining how to do any given integral. Steps 1, 2 and 4 involve nothing more than manipulation of the integrand either through direct manipulation of the integrand or by using a substitution.

The last two steps are simply ideas to think about in going through this strategy. Many students go through this process and concentrate almost exclusively on Step 3 to the exclusion of the other steps.

One very large consequence of that exclusion is that often a simple manipulation or substitution is overlooked that could make the integral very easy to do. Before moving on to the next section we should work a couple of quick problems illustrating a couple of not so obvious simplifications/manipulations and a not so obvious substitution.

Example: Evaluate the following integral.

$$\int \frac{\tan x}{\sec^4 x} dx$$

Solution: This integral almost falls into the form given in 3c. It is a quotient of tangent and secant and we know that sometimes we can use the same methods for products of tangents and secants on quotients.

The process from that section tells us that if we have even powers of secant to strip two of them off and convert the rest to tangents. That won't work here. We can split two secants off, but they would be in the denominator and they won't do us any good there. Remember that the point of splitting them off is so they would be there for the substitution $u = \tan x$.

That requires them to be in the numerator. So, that won't work and so we'll have to find another solution method. There are in fact two solution methods to this integral depending on how you want to go about it. We'll take a look at both.

Solution: In this solution method we could just convert everything to sines and cosines and see if that gives us an integral we can deal with.

$$\int \frac{\tan x}{\sec^4 x} dx = \int \frac{\sin x}{\cos x} \cos^4 x \, dx$$

$$= \int \sin x \cos^3 x \, dx \qquad u = \cos x$$

$$= -\int u^3 du$$

$$= -\frac{1}{4}\cos^4 x + c$$

Note that just converting to sines and cosines won't always work and if it does it won't always work this nicely. Often there will be a lot more work that would need to be done to complete the integral.

Solution: This solution method goes back to dealing with secants and tangents. Let's notice that if we had a secant in the numerator we could just use u = sec x as a substitution and it would be a fairly quick and simple substitution to use. We don't have a secant in the numerator. However we could very easily get a secant in the numerator simply by multiplying the numerator and denominator by secant.

$$\int \frac{\tan x}{\sec^4 x} dx = \int \frac{\tan x \sec x}{\cos^5 x} dx \qquad u = \sec x$$

$$= \int \frac{1}{u^5} du$$

$$= -\frac{1}{4}\frac{1}{\sec^4 x} + c$$

$$= -\frac{1}{4}\cos^4 x + c$$

In the previous example we saw two "simplifications" that allowed us to do the integral. The first was using identities to rewrite the integral into terms we could deal with and the second involved multiplying the numerator and the denominator by something to again put the integral into terms we could deal with.

Using identities to rewrite an integral is an important "simplification" and we should not forget about it. Integrals can often be greatly simplified or at least put into a form that can be dealt with by using an identity. The second "simplification" is not used as often, but does show up on occasion so again, it's best to not forget about it. In fact, let's take another look at an example in which multiplying the numerator and denominator by something will allow us to do an integral.

Example: Evaluate the following integral.

$$\int \frac{1}{1+\sin x}\,dx$$

Solution: This is an integral in which if we just concentrate on the third step we won't get anywhere. This integral doesn't appear to be any of the kinds of integrals that we worked in this chapter.

We can do the integral however, if we do the following,

$$\begin{aligned}\int \frac{1}{1+\sin x}\,dx &= \int \frac{1}{1+\sin x}\,\frac{1-\sin x}{1-\sin x}\,dx \\ &= \int \frac{1-\sin x}{1-\sin^2 x}\,dx\end{aligned}$$

This does not appear to have done anything for us. However, if we now remember the first "simplification" we looked at above we will notice that we can use an identity to rewrite the denominator. Once we do that we can further reduce the integral into something we can deal with.

$$\begin{aligned}\int \frac{1}{1+\sin x}\,dx &= \int \frac{1-\sin x}{\cos^2 x}\,dx \\ &= \int \frac{1}{\cos^2 x} - \frac{\sin x}{\cos x}\,\frac{1}{\cos x}\,dx \\ &= \int \sec^2 x - \tan x \sec x\,dx \\ &= \tan x - \sec x + c\end{aligned}$$

So, we've seen once again that multiplying the numerator and denominator by something can put the integral into a form that we can integrate. Notice as well that this example also showed that "simplifications" do not necessarily put an integral into a simpler form. They only put the integral into a form that is easier to integrate.

Let's now take a quick look at an example of a substitution that is not so obvious.

Example: Evaluate the following integral.

$$\int \cos\left(\sqrt{x}\right)dx$$

Solution: We introduced this example saying that the substitution was not so obvious. However, this is really an integral that falls into the form given by 3e in our strategy above. However, many people miss that form and so don't think about it. So, let's try the following substitution.

$$u = \sqrt{x} \qquad x = u^2 \qquad dx = 2udu$$

With this substitution the integral becomes,

$$\int \cos\left(\sqrt{x}\right)dx = 2\int u \cos udu$$

This is now an integration by parts integral. Remember that often we will need to use more than one technique to completely do the integral. This is a fairly simple integration by parts problem so I'll leave the remainder of the details to you to check.

$$\int \cos\left(\sqrt{x}\right)dx = 2\left(\cos\left(\sqrt{x}\right) + \sqrt{x}\sin\left(\sqrt{x}\right)\right) + c$$

Before leaving this section we should also point out that there are integrals out there in the world that just can't be done in terms of functions that we know. Some examples of these are.

$$\int e^{-x2}dx \qquad \int \cos(x^2)dx \qquad \int \frac{\sin(x)}{x}dx \qquad \int \cos(e^x)dx$$

That doesn't mean that these integrals can't be done at some level. If you go to a computer algebra system such as Maple or Mathematica and have it do these integrals here is what it will return the following.

$$\int e^{-x2}dx = \frac{\sqrt{\pi}}{2}erf(x)$$

$$\int \cos(x^2)dx = \sqrt{\frac{\pi}{2}}\,\text{Fresnel}\,C\left(x\sqrt{\frac{2}{\pi}}\right)$$

$$\int \frac{\sin(x)}{x}dx = \text{Si}(x)$$

$$\int \cos(e^x)dx = Ci(e^x)$$

So it appears that these integrals can in fact be done. However this is a little misleading. Here are the definitions of each of the functions given above.

Error Function

$$erf(x) = \frac{2}{\sqrt{\pi}} \int_0^x e^{-t^2} dt$$

The Sine Integral

$$\text{Si}(x) = \int_0^x \frac{\sin t}{t} dt$$

The Fresnel Cosine Integral

$$\text{Fresnel } C(x) = \int_0^x \cos\left(\frac{\pi}{2} t^2\right) dt$$

The Cosine Integral

$$Ci(x) = \gamma + \ln(x) + \int_0^x \frac{\cos t - 1}{t} dt$$

Where is the Euler-Mascheroni constant.

Note that the first three are simply defined in terms of themselves and so when we say we can integrate them all we are really doing is renaming the integral. The fourth one is a little different and yet it is still defined in terms of an integral that can't be done in practice.

6

Sequences and Series of Functions

TAYLOR'S THEOREM

In many respects, the rational functions are the simplest functions of analysis. They are formed by a finite number of applications of the rational operations of calculation, while in the last resort the formation of every other function involves a more or less concealed passage to the limit from rational functions. The questions whether and how a given function can be expressed approximately by rational functions, in particular, by polynomials are therefore of great importance in theory as well as in practice.

Logarithm and the Inverse Tangent

The Logarithm: We begin with some special cases in which the integration of the geometrical progression leads almost at once to the desired approximations. We first remind the reader of the following fact. For $q \neq 1$ and for positive integers n, we have

$$\frac{1}{1-q} = 1+q+q^2+...+q^{n-1}+r_n,$$

where

$$r_n = \frac{q^n}{1-q}$$

If $q < 1$, the remainder r_n tends to 0 as n increases and we then obtain the infinite geometric series $1 + q + q^2 + ...$ with the sum $\frac{1}{1-q}$

We take as our starting-point the formula,

$$\log(1+x) = \int_0^x \frac{dt}{1+t}$$

and expand the integrand in accordance with the above formula, setting $q = -t$. Then we obtain at once by integration

$$\log(1+x) = x - \frac{x^2}{2} + \frac{x^3}{3} - \frac{x^4}{4} + - \ldots + (-1)^{n-1}\frac{x^n}{n} + R_n,$$

where

$$R_n = \int_0^x r_n dt = (-1)^n \int_0^x \frac{t^n dt}{1+t}$$

Hence, for any positive integer n, we have expressed the function log $(1 + x)$ approximately by a polynomial of the n-th degree, namely $x - \frac{x^2}{2} + \frac{x^3}{3} - + \ldots + (-1)^{n-1}\frac{x^n}{n}$; at the same time, the quantity R_n - the remainder - specifies the size of the error made in the approximation.

In order to estimate the accuracy of this approximation, we need only have an estimate for the remainder R_n; such an estimate is given to us immediately by the integral estimates. If we assume at first that $x \geq 0$, then the integrand is nowhere negative in the entire interval of integration and nowhere exceeds t^n. Consequently,

$$|R_n| \leq \int_0^x t^n dt = \frac{x^{n+1}}{n+1},$$

and we therefore see that, for every value of x in the interval $0 < x < 1$, this remainder can be made as small as we please by choosing n large enough. If, on the other hand, the quantity x is in the interval $-1 < x < 0$, the integrand will not change sign and its absolute value will not exceed $|t|^n/(1 + x)$, and we thus obtain the remainder estimate

$$|R_n| \leq \frac{1}{1+x}\int_0^{|x|} t^n dt = \frac{|x|^{n+1}}{(1+x)(n+1)}$$

Hence, we see that here again the remainder is arbitrarily small when we make n sufficiently large. Of course, our estimate has no meaning for $x = -1$.

Summing up, we can say that

$$\log(1+x) = x - \frac{x^2}{2} + \frac{x^3}{3} - + \ldots + (-1)^{n-1}\frac{x^n}{n} + R_n$$

where the remainder R_n tends to zero as n increases, provided that x lies in the interval $-1 < x < 0$; note that this interval is open on the left hand side and closed on the right hand side. In fact, we can find from the above inequalities an estimate for the remainder, independent of x, which is valid for all values of x in the interval $-1 + h < x < 1$, where h is a number such that $0 < h < 1$. In fact, then $|R_n| \leq \frac{1}{h}\frac{1}{n+1}$ and this formula shows that in the entire interval the function $\log(1 + x)$ is expressed approximately by our polynomial of the n-th degree, the error being nowhere greater than $1/h(n + 1)$.

We leave it to the reader to convince himself that for all values of x for which $|x| > 1$, the remainder not only fails to approach zero, but, in fact, increases numerically beyond all bounds as n increases, so that for such values of x our polynomial does not yield an approximation to the logarithm. The fact that in the above interval the remainder R_n tends to zero may be expressed by saying that we have in this interval for the logarithm the infinite series.

$$\log(1+x) = x - \frac{x^2}{2} + \frac{x^3}{3} - \frac{x^4}{4} + - \ldots$$

If we insert in this series the particular value $x = 1$, we obtain the remarkable formula

$$\log 2 = 1 - \frac{1}{2} + \frac{1}{3} - \frac{1}{4} + - \ldots$$

This is one of the relations the discovery of which left a deep impression on the minds of the first pioneers of the differential and integral calculus. The above approximation for the logarithm leads us to another formula which is useful for many purposes, particularly in numerical calculations. Provided that $-1 < x < 1$, we have only to write $-x$ in place of x in the above formula to obtain,

$$\log(1-x) = -x - \frac{x^2}{2} - \frac{x^3}{3} - \frac{x^4}{4} - \ldots - S_n$$

Assuming that n is even and subtracting, we find

$$\frac{1}{2}\log\frac{1+x}{1-x} = x+\frac{x^3}{3}+\frac{x^5}{5}+\ldots+\frac{x^{n-1}}{n-1}+R_n,$$

where R_n is given by

$$R_n = \frac{1}{2}(R_n+S_n) = \frac{1}{2}\int_0^x t^n\left(\frac{1}{1+t}+\frac{1}{1-t}\right)dt = \int_0^x \frac{t^n}{1-t^2}dt$$

Because $|R_n| \leq \frac{|x^{n+1}|}{n+1}\frac{1}{1-x^2}$, the remainder tends to zero as n increases, a fact which we again express by writing the expansion as an infinite series:

$$\frac{1}{2}\log\frac{1+x}{1-x} = ar\tanh x = x+\frac{x^3}{3}+\frac{x^5}{5}+\frac{x^7}{7}+\ldots,$$

for all values of x such that $|x| < 1$.

An advantage of this formula is that as x traverses the interval -1 to $+1$, the expression $(1 + x)/(1 - x)$ ranges over all positive numbers, whence, if the value of x is chosen suitably, this series enables us to calculate the value of the logarithm of any positive number with an error not exceeding the above estimate for R. The Inverse Tangent: We can treat the inverse tangent in a similar way, if we begin with the formula, true for every positive integer n,

$$\frac{1}{1+t^2} = 1-t^2+t^4-+\ldots+(-1)^{n-1}t^{2n-2}+r_n,$$

where,

$$r_n = (-1)^n \frac{t^{2n}}{1+t^2}$$

By integration, we obtain,

$$\text{ar}\tan x = x-\frac{x^3}{3}+\frac{x^5}{5}-+\ldots+(-1)^{n-1}\frac{x^{2n-1}}{2n-1}+R_n,$$

$$= (-1)^n\int_0^x \frac{t^{2n}}{1+t^2}dt,$$

and see at once that in the interval $-1 \rangle x \rangle 1$ the remainder R_n tends to zero as n increases, because, by the mean value theorem of the integral calculus,

$$|R_n| \leq \int_0^{|x|} t^{2n} dt = \frac{|x|^{2n+1}}{2n+1}$$

Using this formula for the remainder, we can also readily show that for $|x| > 1$ the absolute value of the remainder increases beyond all bounds as n increases. We have thus derived the infinite series

$$\text{arc tan} x = x - \frac{x^3}{3} + \frac{x^5}{5} - + \ldots,$$

valid for $|x|$ 1. For $x = 1$, since arc tan $1 = \pi/4$, we have

$$\frac{\pi}{4} = 1 - \frac{1}{3} + \frac{1}{3} - + \ldots,$$

as remarkable a formula as the one found previously for log 2.

Taylor's Theorem

An approximate representation by rational functions, as in the special cases above, can also be obtained in the case of an arbitrary function $f(x)$ for which we only assume that for all values of the independent variable in an assigned closed interval the function has continuous derivatives at least up to the (n+l)-th order. In most of the cases, which actually occur, the existence and continuity of all the derivatives of the function are known to begin with, so that we can choose for n any arbitrary integer.

The approximation formula, which we shall now derive, was discovered in the early days of the differential and integral calculus by Newton's student Taylor and is known as Taylor's theorem. A special case of this theorem is often referred to, without historical justification, as Maclaurin's theorem. We will not follow this practice.

TAYLOR'S THEOREM FOR POLYNOMIALS

In order to get a clear idea of the problem, we shall consider first the case where $f(x) = a_0 + a_1x + a_2x^2 + \ldots + a_nx^n$ is itself a polynomial of the n-th degree. We can then easily express the coefficients of this polynomial in terms of the derivatives of $f(x)$ at the point x=0.

In fact, differentiating both sides of the equation once, twice, etc., with respect to x and setting $x = 0$, we at once find that the coefficients are,

$$a_0 = f(0),\ a_1 = f'(0),\ a_0 = f''(0), \ldots,\ a_n = \frac{1}{n!} f^{(n)}(0).$$

Any polynomial $f(x)$ of the n-th degree can therefore be written in the form,

$$f(x) = f(0) + xf'(0) + \frac{x^2}{2!} f''(0) + \frac{x^3}{3!} f'''(0) + \ldots + \frac{x^n}{n!} f^{(n)}(0)$$

This formula merely states that the coefficients a_ν can be expressed in terms of the derivatives at $x = 0$ and gives the expressions for them.

We can generalize this Taylor series for the polynomial slightly, if we replace x by $x = \xi + h$ and consider the function $f(\xi) = f(x + h) = g(h)$ as a function of h, for moment thinking of x as fixed and h as the independent variable. It then follows that

$$g'(h) = f'(\xi), \ldots,\ g^{(n)}(h) = f^{(n)}(\xi)$$

and hence, if we set $h = 0$,

$$g'(0) = f'(x), \ldots,\ g^{(n)}(0) = f^{(n)}(x)$$

If we apply the previous formula to the function $f(x + h) = g(h)$, which is itself a polynomial of the n-th degree in h, we immediately obtain the Taylor series

$$f(\xi) = f(x + h) = f(x) + hf'(x) + \frac{h^2}{2!} f''(x)$$

$$+ \frac{h^3}{3!} f'''(x) + \ldots + \frac{h^n}{n!} f^{(n)}(x)$$

TAYLOR'S THEOREM FOR AN ARBITRARY FUNCTION

These formulae suggest that we should seek a similar formula in the case of an arbitrary function $f(x)$, which is not necessarily a polynomial; however, in this case, the formula can lead only to an approximation to the function by a polynomial.

We wish to compare the values of the function f at the point x and at the point $\xi = x+h$, so that $h = \xi - x$. If now n is any positive integer, as a rule, the expression,

$$f(x) + (\xi - x) f'(x) + \ldots + \frac{(\xi - x)^n}{n!} f^{(n)}(x)$$

will not be an exact expression for the functional value $f(\xi)$, whence we must set

$$f(x) = f(x) + (x - x)f'(x) + \frac{(\xi - x)^2}{2!} f''(x)$$

$$+ \ldots \frac{(\xi - x)^n}{n!} f^{(n)}(x) + R_n,$$

where the expression R_n denotes the remainder when $f(\xi)$ is replaced by the expression $f(x) + f'(x)(\xi - x) + \ldots$ In the first instance, this equation is nothing but a formal definition of the expression R_n. Its significance lies in the fact that we can easily find a neat and useful expression for R_n. For this purpose, we think of the quantity ξ as the fixed and the quantity x as the independent variable. The remainder is then the function $R_n(x)$. By the above equation, this function vanishes for $x = \xi$:

$$R_n(x) = 0$$

Moreover, by differentiation, we obtain,

$$R_n'(x) = \frac{(\xi - x)^n}{n!} f^{(n+1)}(x)$$

In fact, if we differentiate this equation for the remainder with respect to x, we obtain 0 on the left hand side, since $f(\xi)$ does not depend on x and is therefore to be regarded as a constant. On the right hand side, we differentiate each term by the rule for products and find that all the terms cancel except the last one, which is written above with a minus sign. Now, by the fundamental theorem of the integral calculus,

$$R_n(x) = R_n(x) - R_n(x)$$

$$= \int_{\xi}^{x} R_n'(t)\,dt = -\int_{x}^{\xi} R_n'(t)\,dt,$$

so that we obtain the formula,

$$R_n(x) = \int_{x}^{x+h} \frac{(x + h - t)^n}{n!} f^{(n+1)}(t)\,dt$$

If we introduce a new integration variable τ by means of the equation $\tau = t - x$, this becomes,

$$R_n = \frac{1}{n!}\int_0^h (h-\tau)^{nf(n+1)}(x+\tau)\,d\tau$$

Collecting these results, we can make the statement:

If the function $f(x)$ has continuous derivatives up to order $(n + 1)$ in the interval under consideration, then

$$f(x+h) = f(x) + hf'(x) + \frac{h^2}{2!}f''(x) + \frac{h^3}{3!}f'''(x)$$

$$+ \dots + \frac{h^n}{n!}f^{(n)}(x) + R_n,$$

or (the equivalent expression for $h = \xi - x$)

$$f(x) = f(x) + (x - x)f'(x) + \frac{(\xi - x)^2}{2!}f''(x)$$

$$+ \dots \frac{(\xi - x)^n}{n!}f^{(n)}(x) + R_n,$$

where the remainder R_n is given by

$$R_n = \frac{1}{n!}\int_0^h (h-\tau)^n f^{(n+1)}(x+\tau)\,d\tau$$

In particular, if we set $x = 0$ and then replace h by x, we obtain

$$f(x) = f(0) + \frac{x}{1!}f'(0) + \frac{x^2}{2!}f''(0) + \dots + \frac{x^n}{n!}f^{(n)}(0) + R_n$$

with the remainder

$$R_n = \frac{1}{n!}\int_0^x (x-\tau)^n f^{(n+1)}(\tau)\,d\tau$$

These formulae are known as Taylor's theorem. They give expressions for the functions $f(x + h)$ and $f(x)$, respectively, in terms of polynomials of degree n in h and in x, respectively (the so-called polynomial of approximation), and a remainder. The polynomial of approximation is characterized by the fact that, when $h = 0$ (or $x = 0$, as the case may be), its value and that of its first

n derivatives are the same as those of the given function and its first n derivatives. In contrast to the Taylor series for polynomials, which do not require a remainder, the remainder and the expression for it are here essential.

The significance of the formula lies in the fact that the remainder, even though it has a more complicated form than the other terms of the formula, nevertheless provides useful means for estimating the accuracy with which the sum of the first $n + 1$ terms,

$$f(0)+\frac{x}{1!}f'(0)+\frac{x^2}{2!}f''(0)+\ldots+\frac{x^n}{n!}f^{(n)}(0),$$

represents the function $f(x)$.

Estimation of the Remainder: Whether the first $n + 1$ terms of Taylor's series actually yield a sufficiently good approximation to the function depends naturally on whether the remainder is sufficiently small.

Hence we turn now to the estimation of this remainder. Such an estimate can most easily be made by means of the mean value theorem of the integral calculus.

We employ this theorem in the form

$$\int_0^h p(\tau)\phi(\tau)d\tau = \phi(\theta h)\int_0^h p(\tau)d\tau,$$

where $p(\tau)$ is a continuous function which is negative nowhere in the interval of integration and $\phi(\tau)$ is merely a continuous function there, while q is a number in the interval $0 < q < 1$. In fact, we may assume that $0 < \theta < 1$, but this is not important here. If we set in the expression for the remainder $p(\tau)=(h-\tau)$, we obtain

$$R_n = \frac{h^{n+1}}{(n+1)!}f^{(n+1)}(x+\theta h),$$

while if we set $p(\tau) = 1$, we obtain

$$R_n = \frac{h^{n+1}}{n!}(1-\theta)^n f^{(n+1)}(x+\theta h),$$

which is less important for us and is stated here only for the sake of completeness. In these formulae, θ denotes a certain number in the interval $0 \} \theta \} 1$, the value of which, in general, cannot be specified more accurately; of course, as a rule, this value is different in the two formulae for the remainder and depends, in addition, on n, x and h. The first form of the remainder was given by Lagrange, the second by Cauchy, whence they received their names. These expressions for the remainder, as well as others, can be derived from the mean value theorem of the differential calculus and from the generalized mean value theorem, respectively.

We apply these theorems to the function $R_n = R_n(x) - R_n(\xi)$ and to the pair of functions $R_n(x)$ and $(x - \xi)^{n+1}$, respectively, where we consider ξ to be fixed and employ the formula

$$R_n'(x) = -\frac{(\xi - x)^n}{n!} f^{(n+1)}(x)$$

These methods of deriving the formulae for the remainder emphasize more the fact that Taylor's theorem is a generalization of the mean value theorem; they also offer the advantage, which for many theoretical purposes is important, that we need only assume the existence and not the continuity of the $(n + 1)$-th derivative.

However, on the other hand, we lose the advantage of having an exact expression for the remainder in the form of an integral. Our interest will be directed chiefly towards discovering whether the remainder R_n tends to zero as n increases; if this is the case, the larger we choose n the more accurately is $f(x + h)$ represented by the corresponding polynomial in h. In this case, we say that we have expanded the function in an infinite Taylor series

$$f(x+h) = f(x) + \frac{h}{1!} f'(x) + \frac{h^2}{2!} f''(x) + \frac{h^3}{3!} f'''(x) + \ldots,$$

or, in particular, if we first set $x = 0$ and then write x in place of h,

$$f(x+h) = f(0) + \frac{x}{1!} f'(0) + \frac{x^2}{2!} f''(0) + \frac{x^3}{3!} f'''(0) + \ldots.$$

We shall encounter examples in the next section. However, first of all, we wish to point out the second important point of view arising from the consideration of Taylor series.

If we think in the first formula of the quantity h as becoming smaller and tending to zero, then, in the terminology of, the various terms of the series will tend to zero at different orders of magnitude; we accordingly call the expression $f(x)$ the term of zero order in the Taylor series, the expression $hf'(x)$ the term of first order, the expression $hf''(x)/2!$ the term of second order, etc. We gather from the form of our remainder the fact: In expanding a function as far as the term of n-th order, we make an error which tends to zero at order $(n+1)$ as $h < 0$.

Many important applications depend on this fact. It shows us that the nearer the point $x + h$ lies to the point x, the better is the representation of the function $f(x+h)$ by the polynomial approximation and that, in a given case, the approximation can be improved in the immediate neighbourhood of the point x by increasing the value of n.

Expansions of the Elementary Functions

We shall now use the general results of the preceding section in order to express the elementary functions approximately by polynomials and to expand them in Taylor series. However, we shall retrict ourselves to those functions for which the coefficients of the expansion in series are given by simple laws of formation. The Exponential function:

The simplest example is offered by the exponential function $f(x) = e^x$, all the derivatives are which identical with $f(x)$ and therefore have the value 1 for $x = 0$. Hence, using Lagrange's form for the remainder.

$$e^{\alpha} = 1 + \frac{x}{1!} + \frac{x^2}{2!} + \frac{x^3}{3!} + \ldots + \frac{x^n}{n!} + \frac{x^{n+1}}{(n+1)!}e^{\theta x}$$

If we now let n increase beyond all bounds, the remainder will tend to zero, no matter what fixed value of x has been chosen, because $|e^{\theta x}| \succ e^{|x|}$ from the start.

We now choose a fixed integer m large than $2|x|$, then for $n \succ m$,

$$\frac{|x|}{n} < \frac{1}{2}; \left|\frac{x^{n+1}}{(n+1)!}\right| = \frac{|x^m|}{m!} \cdot \frac{|x|}{m+1} \cdots \frac{|x|}{n+1} \leqq \frac{|x^m|}{m!}\frac{1}{2^{n+1-m}} \leqq \frac{|2x|^m}{m!}\frac{1}{2^n},$$

$$\text{whence } |R_n| \leqq \frac{|2x|^m}{m!} e^{|x|} \frac{1}{2^n}$$

Since the first two factors on the right hand side are independent of x, while the number $1/2^n$ tends to zero as n increases, our statement is proved. If we think of the number x as not being fixed, but free to vary in the interval $-a < x < a$, where a is a fixed positive number, there follows from the above, if we choose $m > 2a$, that the estimate, $|R_n| \leq \frac{|2a|^m}{m!} e^a \frac{1}{2^n}$ is valid provided $n < m$. Thus, we have specified for the remainder a bound which holds for all values of x in the interval $-a < x < a$ and tends to zero as n. For the function e^x, we can therefore write the expansion as an infinite series

$$e^x = 1 + \frac{x}{1!} + \frac{x^2}{2!} + \frac{x^3}{3!} + \ldots = \sum_{\nu=0}^{\infty} \frac{x^\nu}{\nu!}$$

This expansion is valid for all values of x. Thus, we have again proved that the number e considered is the same as the base of the natural logarithm. Of course, we must employ for numerical calculations the finite form of Taylor's theorem with the remainder; for example, for $x = 1$, this yields

$$e = 1 + 1 + \frac{1}{2!} + \frac{1}{3!} + \ldots + \frac{1}{n!} + \frac{e^\theta}{(n+1)!}$$

If we wish to compute e with an error of at most 1/10,000, we need only choose n so large that the remainder is certainly less than 1/10,000; since this remainder is certainly less* than $3/(n + 1)!$, it is sufficient to choose n = 7, since 8!>30,000. We thus obtain the approximate value $e = 2.71822$ with an error less than 0.0001. We do not take here into account the error due to neglect of the figures in the sixth decimal place. We have used here the fact that $e < 3$. This follows immediately from our series for e, because it is always true that $1/n < 1/2^{n-1}$, whence

$$e < 1 + 1 + \frac{1}{2} + \frac{1}{4} + \ldots = 1 + 1/(1 - \frac{1}{2}) = 3$$

Sin x, cos x, sinh x, cosh x: We find for the functions sin x, cos x,

sinh x and cosh x:

$$
\begin{array}{lclllll}
f(x) & = & \sin x & \cos x & \sinh x & \cosh x, \\
f'(x) & = & \cos x & -\sin x & \cosh x & \sinh x, \\
f''(x) & = & -\sin x & -\cos x & \sinh x & \cosh x, \\
f'''(x) & = & -\cos x & \sin x & \cosh x & \sinh x, \\
f^{iv}(x) & = & \sin x & \cos x & \sinh x & \cosh x
\end{array}
$$

If $f(x) = \sin x$ or $f(x) = \cos x$, the n-th derivative can always be represented by,

$$f^{(\pi)}(x) = f(x + \frac{1}{2}n\pi)$$

Hence, the coefficients of the even powers of x vanish in the polynomial approximations for sin x and sinhx, while the coefficients of the odd powers vanish in those for cos x and cosh x. Thus, in the first case, the $(2n + 1)$-th and the $(2n + 2)$-th polynomial are identical, while, in the second case, the $2n$-th and the $(2n + 1)$-th polynomials are identical. If we employ in each case the higher order ones of these polynomials, we obtain at once Lagrange's forms of the remainders

$$\sin x = x - \frac{x^3}{3!} + \frac{x^5}{5!} - + \ldots + (-1)^n \frac{x^{2n+1}}{(2n+1)!} + (-1)^{n+1} \frac{x^{2n+3}}{(2n+3)!} \cos(\theta x)$$

$$\cos x = 1 - \frac{x^2}{2!} + \frac{x^4}{4!} - + \ldots + (-1)^n \frac{x^{2n}}{(2n)!} + (-1)^{n+1} \frac{x^{2n+2}}{(2n+2)!} \cos(\theta x),$$

$$\sinh x = x + \frac{x^3}{3!} + \frac{x^5}{5!} - + \ldots + \frac{x^{2n+1}}{(2n+1)!} + \frac{x^{2n+3}}{(2n+3)!} \cosh(\theta x),$$

$$\cosh x = 1 + \frac{x^2}{2!} + \frac{x^4}{4!} + \ldots + \frac{x^{2n}}{(2n)!} + \frac{x^{2n+2}}{(2n+2)!} \cosh(\theta x),$$

where, of course, in each of the four formulae θ denotes a different number in the interval $0 < \theta < 1$, a number which, in addition, depends on n and on x.

In these formulae, we can also make the approximation as exact as we please for each value of x, since the remainder tends to 0 as n increases. We thus obtain the four series

$$\sin x =$$

$$x-\frac{x^3}{3!}+\frac{x^5}{5!}-+\ldots=\sum_{\nu=0}^{\infty}(-1)^{\nu}\frac{x^{2\nu}+1}{(2\nu+1)!},$$

cos$x$ =

$$1-\frac{x^2}{2!}+\frac{x^4}{4!}-+\ldots=\sum_{\nu=0}^{\infty}(-1)^{\nu}\frac{x^{2\nu}}{(2\nu)!},$$

sinh$x$ =

$$x+\frac{x^3}{3!}+\frac{x^5}{5!}+\ \ldots=\sum_{\nu=0}^{\infty}\frac{x^{2\nu+1}}{(2\nu+1)!},$$

$$\cosh x = 1+\frac{x^2}{2!}+\frac{x^4}{4!}+\ \ldots=\sum_{\nu=0}^{\infty}\frac{x^{2\nu}}{(2\nu)!}$$

The last two series can also be obtained formally from the series for e^x in accordance with the definitions of the hyperbolic functions.

THE BINOMIAL SERIES

We may pass over the Taylor series for the functions log $(1 + x)$ and artan x, which we have already dealt with directly in equation. However, we must take up the generalization of the binomial theorem for arbitrary indices, which is one of the most fruitful of Newton's mathematical discoveries and which represents one of the most important cases of the expansion in Taylor series. Our objective is the expansion of the function

$$f(x) = (1+x)^{\alpha}$$

in a Taylor series, where $x > -1$ and α is an arbitrary number, positive or negative, rational or irrational. We have chosen the function $(1 + x)^{\alpha}$ instead of x^{α}, because it is not true at the point $x = 0$ that all the derivatives of x^{α} are continuous except in the trivial case of non-negative values of α.

We calculate first the derivatives of $f(x)$:

$$f'(x) = \alpha(1+x)^{\alpha-1},\ f''(x) = \alpha(\alpha-1)(1+x)^{\alpha-2},\ldots,$$

$$f^{(\nu)}(x) = \alpha(\alpha-1)\ldots(\alpha-\nu+1)(1+x)^{\alpha-\nu}$$

In particular, we have for $x = 0$,

$$f'(0) = \alpha,\ f''(0) = \alpha(\alpha-1),\ldots,\ f^{(\nu)}(0)$$

$$= \alpha(\alpha - 1)\ \alpha(\alpha - \nu + 1)$$

Taylor's theorem then yields

$$(1 + x)^{\alpha} = 1 + ax + \frac{\alpha(\alpha-1)}{2!}x^2$$

$$+ \ldots + \frac{\alpha(\alpha-1)(\alpha-2)\ldots(\alpha-n+1)}{n!}x^n + R_n$$

We must still discuss the remainder. This is not very difficult, but nevertheless is not quite as simple as in the cases treated previously. The result, which we mention here in advance, is that whenever $|x| < 1$, the remainder tends to 0, whence the expression $(1 + x)^{\alpha}$ can be expanded in the infinite binomial series

$$(1+x)^{\alpha} = 1 + \frac{\alpha}{1!}x + \frac{\alpha(\alpha-1)}{2!}x^2 + \ldots = \sum_{\nu=0}^{\infty}\binom{\alpha}{\nu}x^{y}$$

where, for the sake brevity, we have introduced the general binomial coefficients

$$\binom{\alpha}{\nu} = \frac{\alpha(\alpha-1)\ldots(\alpha-\nu+1)}{\nu!}(for\,\nu > 0), \binom{\alpha}{0} = 1$$

Geometrical Applications

The behaviour of a function $f(x)$ in the neighbourhood of a point $x = a$ or of a given curve in the neighbourhood of a point can be studied with increased accuracy by means of Tailor's theorem, because this theorem resolves the increment of the function on passing to a neighbouring point $x = a + h$ into a sum of quantities of the first order, second order, etc.

CONTACT OF CURVES

We shall now use this method, in order to investigate the concept of contact of two curves. If at a point, say the point $x = a$, two curves $y = f(x)$ and $y = g(x)$ do not only intersect, but also have a common tangent, we shall say that at this point the curves touch one another or have contact of the first order.

The Taylor expansions of the functions $f(a + h)$ and $g(a + h)$ then have the same terms of zero and first order in h. If at the point $x=a$ the second derivatives of $f(x)$ and $g(x)$ are also equal to each other, we say that the curves have contact of the second order.

In the Taylor expansions, the terms of second order are then also the same and, if we assume that both functions have at least continuous derivatives of the third order, the difference $D(x) = f(x) - g(x)$ can be expressed in the form

$$D(a + h) = \quad f(a + h) - g(a + h) = \frac{h^3}{3!} D'''(a + qh)$$

where the expression $F(h)$ tends to $f''''(a + h)$ - $g'''(a)$ as h tends to zero, whence the difference $D(a + h)$ vanishes to at least the third order in h. We can proceed in this way and consider the general case, when the Taylor series for $f(x)$ and $g(x)$ are the same up to terms of the n-th order, i.e., when

$$f(a) = g(a), f'(a) = g'(a), f''(a)= g''(a),...,f^{(n)}(a) = g^{(n)}(a)$$

We assume here that the $(n + 1)$-th derivatives are also continuous. Under these conditions, we say that at this point the curves have contact of the n-th order. The difference of the two functions will then be

$$f(a + h) - g(a + h) = \frac{h^{n+1}}{(n+1)} f(h)$$

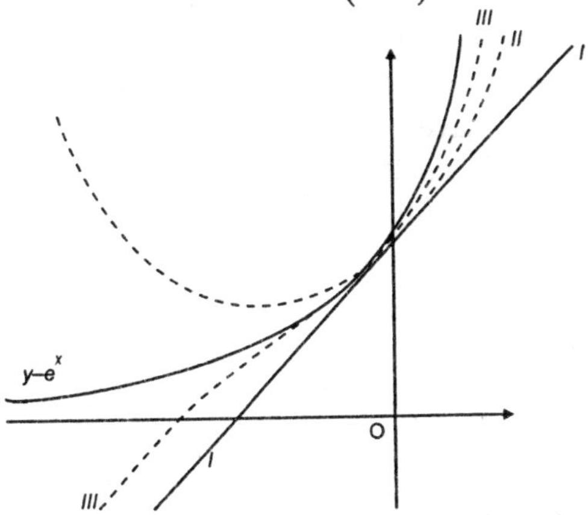

Fig. Osculatin Parabolas of C^2

where, since $0 < \theta < 1$, as h tends to 0, the quantity $F(h) = D^{(n+1)}$

$(a + q\, h)$ tends to $f^{(n+1)}(a)$ - $g^{(n+1)}(a)$. We recognize from this formula that at the point of contact the difference $f(g) - g(x)$ vanishes at least to $(n + 1)$-th order.

The Taylor polynomials are simply defined geometrically by the fact that they are those parabolas of order n which at the given point have contact of the highest possible order with the graph of the given function. Hence they are sometimes called osculating parabolas. For the case $y = e^x$, figures gives the first three osculating parabolas at the point $x = 0$.

If two curves $y = f(x)$ and $y = g(x)$ have contact of order n, the definition does not exclude the possibility that the contact may be of a still higher order, i.e., that also the equation $f^{(n+1)}(a) = g^{(n+1)}(a)$ is true. If this is not the case, i.e., if $f^{(n+1)}(a)$ ýÿ $g^{(n+1)}(a)$, we speak of a contact of exactly the n-th order or say that the order of the contact is exactly n.

The fact that the order of contact of two curves is a genuine geometrical relationship, which is unaffected by a change of axes, is easily verified by means of the formulae for a change of axes. From our formula as well as from our figures, we can at once state a remarkable fact which is often overlooked by beginners. If the contact of two curves is exactly of an even order, that is, if an even number n of derivatives of the two functions have the same value at the point in question, while the $(n+1)$-th derivatives differ, then, in conformity with the above formula, the difference $f(a + h)$ - $g(a + h)$ will have different signs for small positive values and for numerically small negative values of h.

The two curves will then cross at the point of contact. This case occurs, for example, in a contact of the second order, if the third derivatives have different values.

However, if we consider in detail the case of an odd order contact, e.g., the case of an ordinary contact of the first order, the difference $f(a + h)$ - $g(a + h)$ will have the same sign for all numerically small values of h, whether positive or negative; the two curves therefore will not cross in the neighbourhood of the point of contact.

The simplest example of this type is the contact of a curve

with its tangent. The tangent can cross the curve only at points where the contact is at least of second order; it will actually cross the curve at points, where the order of contact is even, e.g., at an ordinary point of inflection, where $f''(x) = 0$, but $f'''(x)$ ýÿ 0. At points, where the order of contact is odd, it will not cross the curve; examples are an ordinary point of the curve where the second derivative is non-zero or the curve $y = x^4$ at the origin.

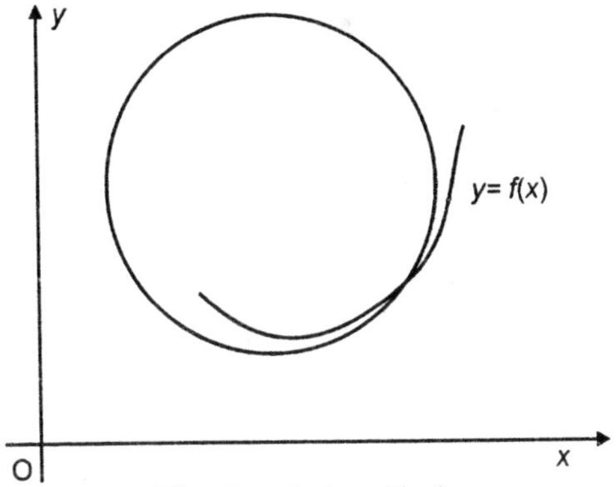

Fig. Osculating Circle

The Circle of Curvature as Osculating Circle:. From this point of view, the concept of the curvature of a curve $y = f(x)$ gains a new intuitive significance. There pass through the definite point of the curve with co-ordinates $x = a$ and $y = b$ an infinite number of circles which touch the curve at the point.

The centres of these circles lie on the normal to the curve and there corresponds to each point of this normal just one such tangent circle. We may expect that we can bring about by a proper choice of the centre of the circle a contact of second order between the curve and the circle.

As a matter of fact, we know from that for the circle of curvature at the point $x = a$, the equation of which is, say, $y = g(x)$, we do not only have $g(a)=f(a)$ and $g'(a) = f'(a)$, but also $g''(a) = f''(a)$, whence the circle of curvature is at the same time the osculating circle at the point of the curve under consideration, i.e., it is the circle which at that point has second order contact with the curve.

In the limiting case of a point of inflection or, in general, of a point at which the curvature is zero and the radius of curvature infinite, the circle of curvature degenerates into the tangent. In ordinary cases, i.e., when the contact at the point in question does not happen to be of an order higher than the second, the circle of curvature will not merely touch the curve, but will also cross it.

ON THE THEORY OF MAXIMA AND MINIMA

A point $x = a$, at which $f'(a) = 0$, has a maximum of the function $f(x)$ if $f''(a)$ is negative, a minimum if $f''(a)$ is positive,whence these conditions are sufficient for the occurrence of a maximum or a minimum. They are by no means necessary, because there are three possibilities in the case where $f''(a) = 0$; at the point in question, the function may have a maximum or a minimum or neither. Examples of the three possibilities are giver by the functions $y=-x^4$, $y=x^4$ and $y = x^3$ at the point $x = 0$.

Taylor's theorem enables us immediately to make a general statement of sufficient conditions for a maximum or a minimum.We merely need expand the function $f(a + h)$ in powers of h; the essential point is then to discover whether the first non-vanishing term contains an even or an odd power of h. In the first case, we have a maximum or a minimum according to whether the coefficient of h is negative or positive; in the second case, we have a horizontal inflectional tangent and neither a maximum nor a minimum. The reader should complete the argument by using the formula for the remainder.

However, the earlier necessary and sufficient condition is more general and more convenient in applications, i.e., provided the first derivative $f'(x)$ vanishes at only a finite number of points; a necessary and sufficient condition for the occurrence of a maximum or minimu,at one of these points is that the first derivative $f'(x)$ changes sign as the point is passed.

INFINITE SERIES

The terms of the infinite series hitherto considered have been constants, whence these series (when convergent) always represented definite numbers. But both in theory and in applications, the series of outstanding importance are those in which the terms are functions of a variable, so that the sum of the series is also a function of the variable, as in the case of Taylor series.

We shall therefore consider a series $g_1(x) + g_2(x) + g_s(x) + \ldots$, in which the functions $g_n(x)$ are defined in an interval $a > x > b$. We will denote the n-th partial sum $g_1(x) + g_2(x) + \ldots + g_n(x)$, by $f_n(x)$. Then the sum $f(x)$ of our series, where it exists, is simply the limit $\lim_{n\to\infty} f_n(x)$.

We may therefore regard the sum of an infinite series of functions as the limit of a sequence of functions $f_1(x), f_2(x), f_n(x), \ldots$ Conversely, for any such sequence of functions $f_1(x), f_2(x), \ldots$, we can form an equivalent series by setting $g_1(x) = f_1(x) \ldots$ and $g_n(x) = f_n(x) - f_{n-1}(x)$ for $n > 1$. Hence, when it is convenient, we can pass from the consideration of series to that of sequences and conversely.

LIMITING PROCESSES WITH FUNCTIONS AND CURVES

We shall now state exactly what we mean by saying that a function $f(x)$ is the limit of a sequence $f_1(x), f_2(x), \ldots, f_n(x)$, in an interval $a \succ x \succ b$. The definition is: The sequence $f_1(x), f_2(x), \ldots$ converges in that interval to the limit function $f(x)$, if at each point x of the interval the values $f_n(x)$ converge in the usual sense to the value $f(x)$. In this case, we shall write $\lim_{n\to\infty} f_n(x) = f(x)$

According to Cauchy's test, we can express the convergence of the sequence without necessarily knowing or stating the limit function $f(x)$. In fact, our sequence of functions will converge to a limit function if, and only if, at each point x in our interval and for every positive number e the quantity $|f_n(x) - f_m(x)|$ is less than e, provided that the numbers n and m are chosen large enough, i.e., larger than a certain number $N = N(\varepsilon)$. This number $N(\varepsilon)$ usually depends on e and x and increases beyond all bounds as ε tends to zero.

We have frequently encountered cases of limits of sequences of functions. We mention only the definition of the power x^a for irrational values of α by the equation

$$x^{\alpha} = \lim_{n\to\infty} x^{r_n}$$

where $r_1, r_2, \ldots, r_n, \ldots$ is a sequence of rational numbers tending to α or the equation

$$e^{\alpha} = \lim_{n\to\infty}\left(1+\frac{x}{n}\right)^{n},$$

where the functions $f_n(x)$ on the right hand side are polynomials of degree n.

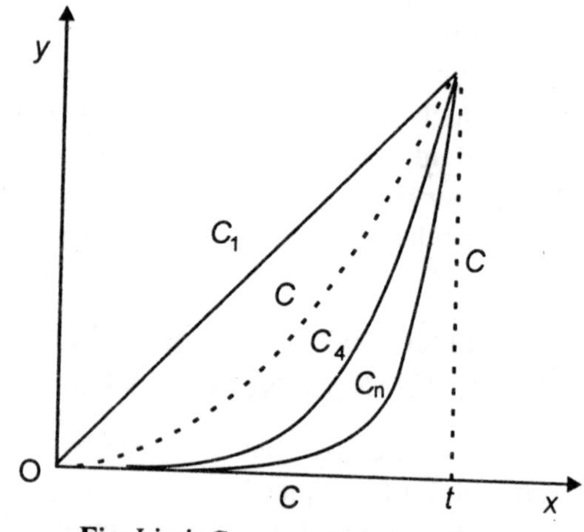

Fig. Limit Curve and Limit Function

The graphical representation of functions by means of curves suggests that we can also speak of limits of sequences of curves, saying, for example, that the graphs of the above limit functions x^a and e^x are to be regarded as the limit curves of the graphs of the functions x^{r_n} *and* $\left(1+\frac{x}{n}\right)^{n}$ respectively. However, there is a fine distinction between the passages to the limit with functions and with curves. Until the middle of the Nineteenth Century, this distinction was not sufficiently recognized and only by having a clear idea of it can we avoid apparent paradoxes. We shall illustrate this point by an example.

Consider the functions

$$f_n(x) = \qquad x^n \ (n = 1, 2, \ldots)$$

in the interval $0 \leq x \leq 1$. All these functions are continuous and the limit function

$$\lim_{n\to\infty} f_n(x) = \qquad f(x)$$

exists. However, this limit function is not continuous. On the contrary, since for all values of n the value of the function $f_n(1) = 1$, the limit $f(1) = 1$,

while, on the other hand, for $0 \leq x < 1$, the limit $f(x) \lim_{n\to\infty} f_n(x) = 0$. The function $f(x)$ is therefore a discontinuous function which at $x=1$ has the value 1 and for all other values of x in the interval has the value 0.

This discontinuity becomes clear, if we consider the graphs C_n of the functions $y = f_n(x)$. These are continuous curves, all of which pass through the origin and the point $x = 1, y = 1$ and which draw in closer and closer to the x-axis as n increases. The curves possess a limit curve C which is not at all discontinuous, but consists of the portion of the x-axis between $x=0$ and $x=1$ and the portion of the line $x=1$ between $y=0$ and $y=1$.

The curves therefore converge to a continuous limit curve with a vertical portion, while the functions converge to a discontinuous limit function. We thus recognize that this discontinuity of the limit function expresses itself by the occurrence in the limit curve of a portion perpendicular to the x-axis. Such a portion must involve a discontinuity in the limit function and, in fact, such a portion is always present when the limit function is discontinuous.

This limit curve is not the graph of the limit function nor can any curve with a vertical portion be the graph of any single-valued function $y = f(x)$; in fact, corresponding to the value of x at which the vertical portion occurs, the curve gives an infinite number of values of y and the function only one. Hence the limit of the graphs of the functions $f_n(x)$ is not the same as the graph of the limit $f(x)$ of these functions.

UNIFORM AND NON-UNIFORM CONVERGENCE

The distinction between the concepts of convergence of functions and of convergence of curves introduces a phenomenon of which it is essential that the student should clearly recognize it. This is the so-called non-uniform convergence of sequences or infinite series of functions. Since it is well known that, as a rule, beginners encounter difficulties here, we should discuss the matter in some detail.

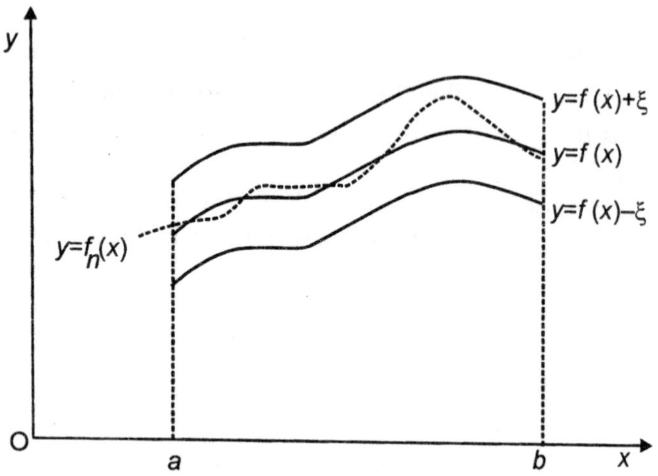

Fig. To Illustrate Uniform Convergence

By definition, a function $f(x)$ is the limit of a sequence $f_1(x)$, $f_2(x)$,... in an interval $a \} x \} b$,only means that the usual limit relationship $f(x) = \lim_{n\to\infty} f(x)$ holds at each point x of the interval. From a naive point of view, one might expect that it would follow automatically from this concept of convergence that if we assign an arbitrary degree of accuracy, say ε=1/1000 or ε =1/100, then from a certain index N onwards all the functions $f_n(x)$ would lie between $f(x) + e$ and $f(x) - \varepsilon$ for all values of x, so that their graphs $y = f_n(x)$ will lie entirely in the strip indicated.

Thus, for every positive ε, there is a corresponding number $N = N(\varepsilon)$, which, of course, will ordinarily increase beyond all bounds as $\varepsilon < 0$, such that, for $n > N$, the difference $|f(x) — f_n(x)| <$

ε, no matter where x is chosen in the interval. If the accuracy of the approximation can be made at least equal to a pre-assigned number ε everywhere in the interval at the same time, i.e., by everywhere choosing the same number $N(\varepsilon)$ independent of x, we may say that the approximation is uniform. One is at first astonished to find that the naive assumption that convergence is necessarily uniform is altogether wrong; in other words, that convergence may very well be non-uniform.

A TEST OF UNIFORM CONVERGENCE

The preceding considerations show that the uniform convergence of a sequence or series is a special property which is not possessed by all sequences and series. We shall now reformulate the concept of uniform convergence.

The convergent series $g_1(x) + g_2(x) + ...$ is said to be uniformly convergent in an interval, if the sum $f(x)$ can be approximated to within ε (where ε is an arbitrarily small positive number) by taking a number of terms which is sufficiently large and which is the same throughout the interval. First of all, we assume that the series $g_1(x) + g_2(x) + ...$ converges at every point of a certain interval to a limit function $f(x)$; we denote by $f_n(x)$ the n-th partial sum of the series $f_n(x) = g_1(x) + g_n(x)$ and by $R_n(x)$ the remainder of the series after n terms:

$$R_n(x) = \qquad f(x) - f_n(x)$$

The series $g_1(x) + g_2(x) + ...$ is said to be uniformly convergent in the interval if there corresponds to every positive number ε a number N, which only depends on e and not on x, such that for $n > N$ one has the inequality $|R(x)|=|f(x)-f_n(x)|<e$ for all values of x *in* the interval.

Expressed more figuratively, the partial sum $f_n(x)$ represents the sum $f(x)$ to within an error of less than e everywhere in the interval at the same time, provided only that n is chosen large enough. By Cauchy's test, we readily see that the series converges uniformly, if, and only if the difference $|f_n(x) - f_m(x)|$ can be made less than an arbitrary quantity e everywhere in the interval by choosing n and m larger than a number N independent of x. In fact, firstly, if the convergence is uniform, we can make both $|f_n(x) - f(x)|$ and $|f_m(x) - f(x)|$ less than $e/2$ by choosing n and m larger than a number N independent of x, whence it follows that $|f_n(x) - f_m(x)| < e$; and, secondly, if $|f_n(x) - f_m(x)| < e$, for all values of x whenever n and m are larger than N, then, on choosing any fixed value of $n > N$ and letting m increase beyond all bounds, we have

$$|f_n(x) - f(x)| = \lim_{m\to\infty} |f_n(x) - f_m(x)| \leqq \varepsilon,$$

for every value of x, whence the convergence is uniform. If we wish to speak of the uniform convergence of a sequence of functions, we have only to make minute changes in the above definition; the sequence $f_1(x), f_2(x), \ldots$ converges uniformly to $f(x)$ in an interval, if the difference $|f(x)-f_n(x)|$ can be made less than ε everywhere in the interval by choosing n larger than a number N independent of x.

As above, a necessary and sufficient condition for the uniform convergence of the sequence is that $|f_n(x) - f_m(x)| < \varepsilon$ for all values of x when both n and m are larger than a certain number N, which depends on ε but not on x.

We shall soon see that it is just this condition of uniform convergence which makes infinite series and other limiting processes with functions into convenient and useful tools of analysis. Fortunately, in the limiting processes, usually encountered in the Calculus and its applications, non-uniform convergence is a sort of exceptional phenomenon which will scarcely trouble us in our present applications of analysis.

In most cases, the uniformity of the convergence of a series is established by means of the criterion:

If the terms of the series $\sum_{\nu=1}^{\infty} g_\nu(X)$ satisfy the condition $|g_\nu(X)| \leqq a_\nu$, where the numbers a_ϖ are constants which form a convergent series $\sum_{\nu=1}^{\infty} a_\nu$ then the series $\sum_{\nu=1}^{\infty} g_\nu(X)$ converges uniformly (and, we incidentally remark, absolutely).

In fact, we have then,

$$\left|\sum_{\nu=n}^{m} g_\nu(x)\right| \leqq \sum_{\nu=n}^{m} \left|g_\nu(x)\right| \leqq \sum_{\nu=n}^{m} a_\nu,$$

and since, by Cauchy's test, the sum $\sum_{\nu=n}^{m} a_\nu$, can be made arbitrarily small by choosing n and $m > n$ large enough, this expresses exactly the necessary and sufficient condition for uniform convergence. A first example is offered by the geometric series $1 + x + x + \ldots$, where x is restricted to the interval $|x| \leq q$, a positive number less than 1. The terms of the series are then numerically less than or equal to the terms of the convergent geometric series $\sum q^\nu$

A further example is the trigonometric series,

$$\frac{c_1 \sin(x-\delta_1)}{1^2} + \frac{c_2 \sin(x-\delta_2)}{2^2} + \frac{c_1 \sin(x-\delta_3)}{3^3} + \ldots,$$

provided that $|c_n| < c$, a positive constant independent of n. Because we have then

$$g_n(x) = \frac{c_n \sin(x-\delta_n)}{n^2}, \text{ so that } |g_n(x)| < \frac{c}{n^2}$$

Hence the uniform and absolute convergence of the trigonometric series follows from the convergence of the series

CONTINUITY OF THE SUM OF A UNIFORMLY CONVERGENT SERIES

As we have already hinted, the significance of the uniform convergence of an infinite series lies in the fact that a uniformly convergent series in many respects behaves exactly like the sum of a finite number of functions. For example, the sum of a finite number of continuous functions is itself continuous and, correspondingly, we have the theorem:

If a series of continuous terms converges uniformly in an interval, its sum is also a continuous function. The proof is quite simple. We subdivide the series $f(x) = g_1(x) + g_2(x) + \ldots$ into the m-th partial sum $f_n(x)$ plus the remainder $R_n(x)$.

As usual,

$$f_n(x) = g_1(x) + ... \; g_n(x).$$

If now any positive number ε is assigned, we can,by virtue of the uniform convergence, choose the number n so large that the remainder is less than $\varepsilon/4$ throughout the entire interval, whence $|R_n(x+h) - R_n(x)| < \frac{\varepsilon}{2}$ for every pair of numbers x and $x + h$ in the interval. The partial sum $f_n(x)$ consists of the sum of a finite number of continuous functions, whence it is continuous; thus, for each point x in the interval, we can choose a positive δ so small that $|f_n(x+h) - f_n(x)| < \frac{\varepsilon}{2}$, provided $|h| < \delta$ and the points x and $m + h$ lie in the interval. It then follows that

$$\begin{aligned} |f(x+h) - f(x)| &= |f_n(x+h) - f_n(x) + R_n(x+h) - R_n(x)| \\ &\leq |f_n(x+h) - f_n(x)| + |R_n(x+h) - R_n(x)| < \varepsilon, \end{aligned}$$

which expresses the continuity of our function. The significance of this theorem becomes clear when we recall that the sums of non-uniformly convergent series of continuous functions are not necessarily continuous, as our previous examples show. We may conclude from the preceding theorem that, if the sum of a convergent series of continuous functions has a point of discontinuity, then in every neighbourhood of this point the convergence is non-uniform, whence every representation of discontinuous functions by series of continuous functions is based on the use of non-uniformly convergent limiting processes.

INTEGRATION OF UNIFORMLY CONVERGENT SERIES

A sum of a finite number of continuous functions can be integrated term by term, i.e., the integral of the sum can be found by integrating each term separately and adding the integrals. In the case of a convergent infinite series, the same procedure is possible, provided that the series converges uniformly in the interval of integration.

A series $\sum_{\nu=1}^{\infty} g_\nu(X) = f(X)$ which converges uniformly in an interval can be integrated term by term in that interval, or, more precisely, if a and x are two numbers in the interval of uniform convergence, the series,

$$\sum_{\nu=1}^{\infty} \int_a^X g_\nu(t)\,dt$$

converges and, in fact, converges uniformly with respect to x for each fixed value of a, its sum being equal to $\int_a^X f(t)\,dt.$.

In order to prove this we write, as before,

$$f(x) = \qquad \sum_{\nu=1}^{\infty} g_\nu(x) = f_n(x) + R_n(x)$$

We have assumed that the separate terms of the series are continuous, the sum is also continuous and therefore integrable. Now, if ε is any positive number, we can find a number N so large that, for every $n > N$, one has $|R_n(x)| < \varepsilon$ for every value of x in the interval. By the first mean value theorem of the integral calculus, we have,

$$\left| \int_a^x \{f(t) - f_n(t)\}\,dt \right| \leqq \in l,$$

where l is the length of the interval of integration. Since the integration of the finite sum $f_n(x)$ can be performed term by term, this yields,

$$\left| \int_a^x f(t)\,dt - \sum_1^n \int_a^x g_\nu(t)\,dt \right| \leqq \in l$$

However, since $\varepsilon\, l$ can be made as small as we please, this yields $\sum_{\nu=1}^{\infty} \int_a^x g_\nu(t)\,dt = \lim_{n\to\infty} \sum_{\nu=1}^{n} \int_a^x g_\nu(t)\,dt = \int_a^x f(t)\,dt$, as was to be proved. If we wish to conside instead of infinite series sequences of functions, our result can be expressed as follows: If in an interval the sequence of functions $f_1(x)$, $f_2(x)$, ... tends uniformly to the limit function $f(x)$, then $\int_a^b f(x)\,dx = \lim_{n\to\infty} \int_a^b f_n(x)\,dx$ for every pair of numbers a and b in the interval; in other words, we can then interchange the order of the operations of integration and going to the limit.

This fact is far from being a triviality. It is true that, from a naive point of view such as prevailed in the Eighteenth Century, the interchangeabi]ity of the two processes is hardly to be doubted; however, shows that in the case of non-uniform convergence the above equation might not hold. We need only consider Example 2, in which the integral of the limit function is 0, while the integral of the function $f_n(x)$ over the interval $0 < x < 1$, i.e., the area of the triangle has the value $\int_a^b f_n(x)\,dx = n^{\alpha-2}$, and, when $\alpha > 2$, this does not tend to zero.

We see here immediately from figure that the reason for the difference between $\int_0^1 f_n(x)\,dx = n^{\alpha-2}$, and $\lim_{n\to\infty} \int_0^1 f_n(x)\,dx$ lies in the non-uniformity of the convergence.

On the other hand, by considering values of α in the interval $1 < \alpha < 2$, we see that the equation $\lim_{n\to\infty} \int_0^1 f_n(x)\,dx = \int_0^1 f(x)\,dx$ can be valid, although the convergence is non-uniform.

As a further example, we can integrate term by term the series $\sum_0^\infty g_n(x)$,, where $g_n(x) = x^n - x^{n-1}$ for $n \geq 1$ and $g_0(x) = 1$, between 0 and 1, even though it does not converge uniformly.

Thus, while uniformity of convergence is a sufficient condition for term-by-term integrability, it is by no means a necessary condition. Neglect of this aspect may easily lead to a misunderstanding.

DIFFERENTIATION OF INFINITE SERIES

The behaviour of uniformly converging series or sequences with respect to differentiation is quite different from that with respect to integration. For example, the sequence of functions $f_n(x) = \dfrac{\sin n^2\pi}{n}$ certainly converges uniformly to the limit function $f(x) = 0$, but the derivative $f_n'(x) = n \cos n^2 x$ does not converge everywhere to the derivative of the limit function $f'(0) = 0$, as we see by considering $x = 0$.

In spite of the uniformity of the convergence, we cannot therefore change the order of the process of differentiation and pass on to the limit. Naturally, corresponding statements hold for infinite series. For example, the series,

$$\sin x + \frac{\sin 2^4 x}{2^2} + \frac{\sin 3^4 x}{3^2} + \ldots$$

is absolutely and uniformly convergent, because its terms are numerically not greater than the terms of the convergent series,

$$\frac{1}{1^2} + \frac{1}{2^2} + \frac{1}{3^2} + \ldots.$$

However, if we differentiate the series term by term, we obtain the series $\cos x + 2^2 \cos 2^4 x + 3^2 \cos 3^4 x + \ldots$, which obviously does not converge everywhere, since, for example, it diverges at $x = 0$. The only useful criterion which assures us in special cases that term-by-term differentiation is permissible is given by the theorem:

If, on differentiating a convergent infinite series,

$$\sum_{\nu=0}^{\infty} G_\nu(X) = F(X)$$

term by term, we obtain a uniformly convergent series of continuous terms,

$$\sum_{\nu=0}^{\infty} g_\nu(X) = f(X),$$

then the sum of this last series is equal to the derivative of the sum of the first series. Hence, this theorem expressly requires that, after differentiating the series term by term, we must still investigate whether the result of the differentiation is a uniformly convergent series or not.

The proof of the theorem is almost trivial, because, we can integrate term by term the series obtained by differentiation. Recalling that $g_n(t) = G_n'(t)$, we obtain

$$\int_a^x f(t)\,dt = \int_a^x \left(\sum_{\nu=0}^{\infty} g_\nu(t) \right) dt =$$

$$\sum_{\nu=0}^{\infty}\int_a^x g_\nu(t)\,dt=\sum_{\nu=0}^{\infty}\left(G_\nu(x)-G_\nu(a)\right)$$

$$= F(x) - F(a)$$

Since this is true for every value of x in the interval of uniform convergence, it follows that $f(x) = f(a)$ as was to be proved.

POWER SERIES

Among infinite series, power series occupy the main position. We mean by a power series a series of the type

$$P(x) = c_0 + c_1x + c_2x^2 + \ldots = \sum_{\nu=0}^{\infty} c_\nu x^\nu$$

(a power series in x) or, more generally,

$$P(x) = c_0 + c_1x(x - x_0) + c_2(x - x_0)^2 + \ldots = \sum_{\nu=0}^{\infty} c_\nu (x - x_0)^\nu$$

(a power series in $(x - x_0)$), where x_0 is a fixed number. If we introduce into the last series $\xi = x - x_0$ as a new variable, it becomes a power series $\sum_{\nu=0}^{\infty} c_\nu \xi^\nu$ in the new variable ξ, whence, without any loss of generality, we can confine our attention to power series of the more special form $\sum_{\nu=0}^{\infty} c_\nu x^\nu$

We have considered the approximate representation of functions by polynomials and were thus led to the expansion of functions in Taylor series, which, in fact, are power series. We shall now study power series in somewhat greater detail and obtain expansions of the most important functions in series in simpler and more convenient ways than before.

CONVERGENCE PROPERTIES OF POWER SERIES

There are power series which converge for no value of x, except, of course, for $x = 0$, e.g., the series $x + 2^2x^2 + 3^3x^3 + \ldots + n^nx^n + \ldots$ In fact, if $x < 0$, we can find an integer N such that $|x| > 1/N$. Then the absolute values of all the terms n^nx^n, for which $n >$

N, will be larger than 1 and, as n increases, $nn^n x^n$ will increase beyond all bounds, so that the series fails to converge. On the other hand, there are series which converge for every value of x; for example, the power series for the exponential function,

$$e^x = 1 + x + \frac{x^2}{2!} + \frac{x^3}{3!} + \ldots,$$

the convergence of which for every value of x follows at once from the ratio test. The $(n + 1)$-th term, divided by the n-th term, yields x/n and, whatever number x is chosen, this ratio tends to zero as n increases. The behaviour of power series with regard to convergence is expressed in the fundamental theorem: If a power series in x converges for a value $x = \xi$, it converges absolutely for every value x such that $|x| < |\xi|$ and the convergence is uniform in every interval $|x| \leq \eta$, where η is any positive number less than $|\xi|$. Hence h may lie as close to $|\xi|$ as we please.

The proof is simple. If the series $\sum_{\nu=0}^{\infty} c_\nu \xi^\nu$ converges, its terms tend to 0 as n increases. From this follows the weaker statement that all the terms lie below a bound M independent of n, i.e., $|c_n \xi^\nu| < M$. If now q is any number such that $0<q<1$ and we restrict x to the interval $|x| \leq q|\xi|$, then $|c_n x^\nu| < |c_n x^\nu| q^\nu < Mq^\nu$. Hence, in this interval, the terms of the series are smaller in absolute value than the terms of the convergent geometric series ΣMq^ν, whence follows, by the theorem, the absolute and uniform convergence of the series in the interval $-q|\xi| < x < q|x|$. If a power series does not converge everywhere, i.e., if there is a value $x = x$ for which it diverges, it must diverge for every value of x such that $|x| > |\xi|$. In fact, if it were convergent for such a value of x, by the theorem above, it would have to converge for any numerically smaller value of ξ. Hence, we recognize that a power series, which converges for at least one value of x other than 0 and which diverges for at least one value of x, has an interval of convergence, i.e., there exists a definite positive number r such that the series diverges for $|x| > \rho$ and converges for $|x| < \rho$. No general statement can be made for $|x| = \rho$. The limiting cases, in which the series converges only for x=0 and that which converges everywhere, are expressed symbolically by writing $\rho = 0$ and $\rho = \infty$, respectively.

It is possible to find this interval of convergence directly from the coefficients c_ν of the series. If the limit $\lim_{n\to\infty} \sqrt[n]{|c_n|}$ exists, then

$$\rho = \frac{1}{\lim_{n\to\infty} \sqrt[n]{|c_n|}}$$

In general, ρ is given by the formula

$$\rho = \frac{1}{\overline{\lim}_{n\to\infty} \sqrt[n]{|c_n|}}$$

where $\overline{\lim}$ is the symbol for the upper limit.

For example, $\rho = 1$ for the geometric series $1 + x + x + \ldots$; at the end-points of the interval of convergence, the series diverges. Similarly, for the series for the inverse tangent,

$$\text{arc tan } x = x - x^3/3 + x^5/5 - + \ldots,$$

we have $\rho = 1$ and at both of the interval's end-points $x = 1$ the series converges, as we recognize at once from Leibnitz's test.

We deduce from the uniform convergence the important fact that within its interval of convergence (if such an interval exists) a power series represents a continuous function. Integration and Differentiation of Power Series: On account of the uniformity of convergence, it is always permissible to integrate a power series

$$f(x) = \sum_{\nu=0}^{\infty} c_\nu x^\nu$$

term by term over any closed interval lying entirely within the interval of convergence. We thus obtain the function

$$F(x) = c + \sum_{\nu=0}^{\infty} \frac{c_\nu}{\nu+1} x^{\nu+1},$$

for which

$$F'(x) = f(x)$$

Moreover, since $|c_\nu/(n+1)| \leq |c_\nu|$ for all values of ν, the series obtained by integration converges more rapidly than the original series. We can also differentiate a power series term by term within its interval of convergence, thus obtaining the equation,

$$f'(x) = \sum_{\nu=1}^{\infty} \nu c_\nu x^{\nu-1}$$

In order to prove this statement, we need only show that the series on the right hand side converges uniformly, if x is restricted to an interval lying entirely within the interval of convergence. Now, let x be a number, lying as close to r as we please, for which

$$\sum_{\nu=1}^{\infty} c_\nu \xi^\nu$$

converges; then, as we have seen before, all the numbers $|c_n \xi_n|$ lie below a bound M independently of ν, so that,

$$\left|c_\nu \xi^{\nu-1}\right| < \frac{M}{|\xi|} = N$$

Now let q be any number such that $0 < q < 1$; if we restrict x to the interval $|x| \nleq q|\xi|$, the terms of the series under discussion are not larger than those of the series,

$$\sum_{\nu=1}^{\infty} \left|\nu c_\nu q^{\nu-1}\, \xi^{\nu-1}\right|,$$

and therefore less than those of the series,

$$\sum_{\nu=1} N\nu q^{\nu-1}$$

However, in this last series, the ratio of the $(n + 1)$-th term to the n-th term is $\{(n+1)/n\}q$, which tends to q as n increases. Since $0 < q < 1$, it follows by Criterion IIIa that this series converges. Hence the series, obtained by differentiation, converges uniformly and by the theorem at the end of last section represents the derivative $f'_n(x)$ of the function $f(x)$, which proves our statement.

If we apply this result again to the power series,

$$f'(x) = \sum_{\nu=1}^{\infty} \nu c_\nu x^{\nu-1},$$

we find on differentiating term by term that,

$$f''(x) = \sum_{\nu=2}^{\infty} \nu(\nu-1) c_\nu x^{\nu-2}$$

and, by continuing the process, we arrive at the theorem: Every function represented by a power series can be differentiated as often as we please within the interval of convergence and the differentiation can be performed term by term. As an explicit expression for the k-th derivative, we obtain,

$$f^{(k)}(x) = \sum_{\nu=k}^{\infty} \nu(\nu-1)\ldots(\nu-k+1)c_\nu x^{\nu-k},$$

or in a slightly different form

$$\frac{f^{(k)}(x)}{k!} = \sum_{\nu=k}^{\infty} \binom{\nu}{k} c_\nu x^{\nu-k} = \sum_{\nu=0}^{\infty} \binom{k+\nu}{k} c_{k+\nu} x^\nu$$

These two formulae are frequently useful.

OPERATIONS WITH POWER SERIES

The preceding theorems on the behaviour of power series are our justification for operating in the same way with power series as with polynomials. Obviously, two power series can be added or subtracted by adding or subtracting the corresponding coefficients. It is also clear that a power series, like any other convergent series, can be multiplied by a constant factor by multiplying each term by that factor. We merely mention here without a proof that two power series,

$$f(x) = \sum_{\nu=0}^{\infty} a_\nu x^\nu \; and \; g(x) = \sum_{\nu=0}^{\infty} b_\nu x^\nu$$

can be multiplied by each other like polynomials. In order to be specific, we have the theorem: Throughout the common part of the interval of convergence of two power series, their product is given by the convergent power series,

$$\sum_{\nu=0}^{\infty} c_\nu x^\nu$$

with the coefficients

$$\begin{aligned} c_0 &= a_0 b_0, \\ c_1 &= a_0 b_1 + a_1 b_0, \\ c_2 &= a_0 b_2 + a_1 b_1 + a_2 b_0, \\ &\ldots\ldots\ldots\ldots \\ c_n &= a_0 b_n + a_1 b_{n-1} + \ldots + a_n b_0, \\ &\ldots\ldots\ldots\ldots \end{aligned}$$

Theorem of Uniqueness of Power Series: In the theory of power series, the following fact is important:

If both power seriesm,

$$\sum_{\nu=0}^{\infty} a_\nu x^\nu \, and \sum_{\nu=0}^{\infty} b_\nu x^\nu$$

converge in an interval which contains the point $x = 0$ inside and if in that interval the two series represent the same function $f(x)$, then they are identical, that is, the equation $a_n = b_n$ is true for every value of n. In other words: If at all, a function $f(x)$ can be represented by a power series in x in only one way. Briefly speaking, the representation of a function by a power series is unique. In order to prove this, we need only note that the difference of the two power series, i.e., the power series,

$$\phi(x) = \sum_{\nu=0}^{\infty} c_\nu x^\nu$$

with the coefficients $c_\nu = a_\nu - b_\nu$ represents the function $\phi(x) = f(x) - f(x) = 0$ in the interval, i.e., this last power series converges to the limit 0 everywhere in the interval. In particular, for $x = 0$, the sum of the series must be 0, i.e., $c_0 = 0$, so that $a_0 = b_0$.

We now differentiate the series in the interior of the interval and obtain,

$$\phi'(x) = \sum_{\nu=1}^{\infty} \nu c_\nu x^{\nu-1}$$

However, $\phi'(x)$ is also 0 throughout the interval, whence, in particular, we have for $x = 0$: $c_1 = 0$ or $a_1 = b_1$. Continuing this process of differentiation and then setting $x = 0$, we find successively that all the coefficients c_n are equal to zero, which proves the theorem.

Moreover, we see that we can draw from the above discussion the conclusion: If we take the ν-th derivative of a series $f(x) = \sum a_\nu x^\nu$ and then set $x = 0$, we at once obtain

$$a_\nu = \frac{1}{\nu!} f^{(\nu)}(0)$$

i.e., Every power series which converges for points other than $x = 0$ is the Taylor series of the function which it represents. The uniqueness of the expansion is here expressed by the fact that the coefficients are uniquely determined by the function itself.

EXPANSION OF GIVEN FUNCTIONS IN POWER SERIES

Within its interval of convergence, every power series represents a continuous function with continuous derivatives of all orders. We shall now discuss the converse problem of the expansion of a given function in a power series. In theory, we can always do this by means of Taylor's theorem, however, in practice, we often encounter difficulties with the actual calculation of the n-th derivative and with the estimation of the remainder. But we can often reach our goal more simply by means of the device:

We first write down the statement,

$$f(x) = \sum_{\nu=0}^{\infty} c_\nu x^\nu,$$

where, to begin with, the coefficients c_ν are unknown. Then we determine by some known property of the function $f(x)$ the coefficients and prove subsequently the convergence of the series. The series represents a function and there remains only to prove that this function is identical to $f(x)$. Due to the uniqueness of the expansion in power series, we know that no other series than the one just found can be the required expansion.

We shall now consider some examples of the use of this method. In fact, we have already obtained the series for artan x and log $(1+x)$ by a method, which forms part of the range of ideas. We simply integrated term by term the series for the derivatives of these functions, which we knew to be geometric series.

The Exponential Function: Our problem is to find a function for which $f'(x) = f(x)$ and $f(0) = 1$. If we write down the series with undetermined coefficients $f(x) = c_0 + c_1x + c_2x^2 + \ldots$, and differentiate it, we obtain $f'(x) = c_1 + 2c_2x + 3c_3x^2 + \ldots$.

Since, by assumption, these two power series must be identical, we have the equation $nc_n = c_{n-1}$, which is true for all values of $n \geq 1$. If we observe that, due to the relation $f(0)=1$, the coefficient c_0 must have the value 1, we can compute all the coefficients successively and obtain the power series,

$$f(x) = 1 + \frac{x}{1!} + \frac{x^2}{2!} + \frac{x^3}{3!} + \ldots$$

As is easily shown by the ratio test, this series converges for all values of x and therefore represents a function for which the relations $f'(x) = f(x), f(0) = 1$ are actually fulfilled.

Now, the function e^x certainly possesses these properties; we readily deduce that the function $\phi(x)$ is identical to e^x. In fact, if we form the quotient $f(x)=f(x)/e^x$ and differentiate it, we find

$$\varphi(\xi) = \frac{e^{\alpha} f'(x) - e^{\alpha} f'(x)}{e^{2x}} = 0$$

Hence, the function $\phi(x)$ is a constant and since it has the value 1 for $x = 0$, it must be identically equal to 1, thus proving that our power series and the exponential function are identical.

THE BINOMIAL SERIES

We can now return to the binomial series, this time using the method of undetermined coefficients. We wish to expand the function $f(x) = (1+x)^{\alpha}$ in a power series, whence we write $f(x) = (1 + x)^{\alpha} = c_0 + c_1x + c_2x^2 + \ldots$, where the coefficients c_n are undetermined. We now note that obviously our function satisfies the relation,

$$(1 + x) = f'(x) = \alpha f(x) = \sum_{\nu=0}^{\infty} \alpha c_{\nu} x^{\nu}$$

On the other hand, if we differentiate the series for $f(x)$ term by term and multiply by $(1+ x)$, we obtain $(1 + x) f'(x) = c_1 + (2c_2 + c_1) x + (3c_3 + 2c_2) x^2 + \ldots$; since these two power series for $f'(x)$ must be identical, we have $\alpha c_0 = c_1$, $\alpha c_1 = 2c_2 + c_1$, $\alpha c_2 = 3c_3 + 2c_2, \ldots$

It is now certain that $c_0 = 1$, since our series must have the value 1 for $x = 0$; we thus find in succession the expressions,

$$c_1 = \alpha,\ c_1 = \frac{(\alpha-1)\alpha}{2},\ c_1 = \frac{(\alpha-2)(\alpha-1)\alpha}{3\cdot 2},\ldots,$$

for the coefficients, and, in general, as is readily established,

$$c_\nu = \frac{(\alpha-\nu-1)(\alpha-\nu+2)\ldots(\alpha-1)\alpha}{\nu(\nu-1)\ldots 2.1} = \binom{\alpha}{\nu}$$

Substituting these values for the coefficients, we have the series,

$$\sum_{\nu=0}^{\infty}\binom{\alpha}{\nu}x^{\nu};$$

we must yet investigate the convergence of this series and show that it actually represents $(1 + x)^n$.

By the ratio test, we find that, when α is not a positive integer, the series converges if $|x| < 1$ and diverges if $|x| > 1$; in fact, then the ratio of the $(n + 1)$-th term to the n-th term is $\{(\alpha - n + 1)/n\}x$ and the absolute value of this expression tends to $|x|$ as n increases beyond all bounds. We state here without proof the exact conditions under which this series converges.

If the index a is an integer, the series terminates and is therefore valid for all values of x (as it becomes the ordinary binomial theorem). For all other values of α, the series is absolutely convergent for $|x|<1$ and diverges for $|x|>1$. For $x = +1$, the series converges absolutely if $\alpha > 0$, conditionally if $-1 < \alpha < 0$, and it diverges if $\alpha = -1$. Finally, at $x = -1$, the series is absolutely convergent if $a > 0$, divergent if $a < 0$. Hence, if $|x| < 1$, our series represents a function $f(x)$ which satisfies the condition $(1 + x)f'(x) = af(x)$, as follows from the method of forming the coefficients. Moreover, $f(0) = 1$. But these two conditions ensure that the function $f(x)$ is identical to $(1 + x)^\alpha$. In fact, setting

$$\phi(x) = f(x)/(1 + x)^\alpha,$$

we find that

$$\phi(x) = \frac{(1+x)^\alpha f'(x) - \alpha(1+x)^{\alpha-1}f(x)}{(1+x)^{2\alpha}}$$

hence $f(x)$ is a constant and is always equal to 1, since $\phi(0) = 1$. We have thus proved that for $|x| < 1$,

$$(1+x)^{\alpha}=\sum_{\nu=0}^{\infty}\binom{\alpha}{\nu}x^{\nu}$$

which is the binomial series. We present here the following special cases of the binomial series: The geometric series:

$$\frac{1}{1+x} = (1+x)^{-1}=1-x+x^2-x^3+x^4-+\ldots=\sum_{\nu=0}^{\infty}(-1)^{\nu}x^{\nu};$$

the series $\frac{1}{(1+x)^2} = (1+x)^{-2} = 1 - 2x + 3x^2 - 4x^3 + -\ldots$

$=\sum_{\nu=0}^{\infty}(-1)^{\nu}(\nu+1)x^{\nu}$, which may otherwise be obtained from the geometric series by differentiation and the

$$\sqrt{(1+x)} = (1+x)^{\frac{1}{2}}$$

$$= 1+\frac{1}{2}x-\frac{1}{2.4}x^2+\frac{1.3}{2.4.6}x^3-\frac{1.3.5}{2.4.6.8}x^4+-\ldots,\ \frac{1}{\sqrt{(1+x)}}$$

$$= (1+x)^{-\frac{1}{2}} = 1+\frac{1}{2}x+\frac{1.3}{2.4}x^2-\frac{1.3.5}{2.4.6}x^3\,\frac{1.3.5.7}{2.4.6.8}x^4-+\ldots,$$

series the first two or three terms of which are useful expressions.

The Series for arsin x: This series can be established readily by expansion of the expression $1/\sqrt{(1-t^2)}$ as a binomial series

$$(1-t^2)^{-\frac{1}{2}} = 1+\frac{1}{2}t^2+\frac{1.3}{2.4}t^4+\ldots$$

This series converges if $|t| \rangle 1$, whence it converges uniformly for $|t| < q < 1$. On integrating term by term between 0 and x, we obtain

$$\text{arc}\sin x = x+\frac{1}{2}\frac{x^3}{3}+\frac{1.3}{2.4}\frac{x^5}{5}+\ldots;$$

by the ratio test, we find that this series converges if $|x| < 1$ and diverges if $|x|>1$.

The derivation of this series by Taylor's theorem would be decidedly less convenient owing to the difficulty of estimating the remainder.

The Series for arsinh $x = \log\{x + (1 + x)\}$: We obtain this expansion by a similar method. By the binomial theorem, we write down the series for the derivative of arsinh x:

$$\frac{1}{\sqrt{1+x^2}} =$$

$$1-\frac{1}{2}x^2+\frac{1.3}{2.4}x^4-\frac{1.3.5}{2.4.6}x^6+-...,$$

and then integrate term by term to obtain the expansion

$$\text{arc sinh}x = \qquad x-\frac{1}{2}\frac{x^3}{3}+\frac{1.3}{2.4}\frac{x^5}{5}-+...;$$

with the interval of convergence $-1 < x < 1$.

POWER SERIES WITH COMPLEX TERMS

COMPLEX TERMS INTO POWER SERIES

The similarity between certain power series representing functions which are apparently unrelated led Euler to set up a purely formal link between them, found by giving complex values, in particular, pure imaginary values to the variable x. We shall first do this formally, unhindered by questions of rigour and investigate the results of the process.

The first striking relation of this sort is obtained, if we replace the quantity x in the series for e^x by a pure imaginary $i\phi$, where ϕ is a real number. If we recall the fundamental equation for the imaginary unit i, that is, $i = -1$, from which it follows that $i^3 = -i$, $i^4 = 1$, $i^5 = i$,..., then, on separating the real and imaginary terms of the series, we obtain

$$e^{1\phi} =$$

$$\left(1-\frac{\phi^2}{2!}+\frac{\phi^4}{4!}+\frac{\phi^6}{6!}+-...\right)+i\left(\phi-\frac{\phi^3}{3!}+\frac{\phi^5}{5!}-\frac{\phi^7}{7!}+-...\right),$$

or, in another form, $e^{1\phi} = \cos\phi + i\sin\phi$

This is the well-known and important Euler formula. which is as yet purely formal. It is consistent with De Moivre's theorem which is expressed by the equation $(\cos\phi + i\sin\phi)(\cos\psi + i\sin\psi) = (\cos\phi + \psi) + i\sin(\phi + \psi)$.

By virtue of Euler's formula, this equation merely states that the relation $e^x \cdot e^v = e^{x+v}$ continues to hold for pure imaginary values $x = i\phi$, $y = i\psi$. If we replace the variable x in the power series for $\cos x$ by the pure imaginary ix, we obtain at once the series for $\cosh x$; this relation can be expressed by the equation

$$\cosh x = \cos ix.$$

In the same way, we obtain

$$\sinh x = \frac{1}{i}\sinh ix.$$

Since Euler's formula also yields

$$e^{-if} = \cos\phi - i\sin\phi,$$

we arrive at the exponential expressions for the trigonometric functions

$$\sin x = \frac{e^{ix} - e^{-ix}}{2i}, \ \cos x = \frac{e^{ix} + e^{-ix}}{2}$$

These are exactly analogous to the exponential expressions for the hyperbolic functions and are, in fact, transformed into them by the relations

$$\cos hx = \cos ix, \ \sin h\,x = \frac{1}{i}\sin ix$$

Naturally, corresponding formal relations can also be obtained for the functions $\tan x$, $\tanh x$, $\cot x$ and $\coth x$, which are interrelated by the equations

$$\tanh x = \frac{1}{i}\tan ix, \ \coth x = i\cot ix$$

Finally, similar relations can also be found for the inverse trigonometric and hyperbolic functions. For example, we immediately find from

$$y = \tan x = \frac{e^{ix} - e^{-ix}}{i\left(e^{ix} + e^{-ix}\right)} = \frac{e^{2ix} - 1}{i\left(e^{2ix} + 1\right)}$$

that

$$e^{2ix} = \frac{1 + iy}{1 - iy}$$

It we take the logarithms of both sides of this equation and then write x instead of y and artan x instead of x, we obtain the equation,

$$\text{artan}x = \frac{1}{2i}\log\frac{1+ix}{1-ix},$$

which expresses a remarkable connection between the inverse tangent and the logarithm. If we replace in the known power series for $\log(1+x)/(1-x)$ x by ix, we actually obtain the power series for artan x:

$$\text{artan}x = \frac{1}{i}\left(ix + \frac{(ix)^3}{3} + \frac{(ix)^5}{5} + \ldots\right) = x - \frac{x^3}{3} + \frac{x^5}{5} - + \ldots$$

The above relations have as yet a purely formal character and naturally demand a more exact statement as to the meaning which they are intended to convey. In the next sub-section, we shall indicate how this can be given with the help of function theory.

However, we shall later on only require Euler's formula $e^{i\phi} = \cos\phi + i\sin\phi$. This being so, we can avoid here a thorough analysis. We need only regard the symbol $e^{i\phi}$ as a formal abbreviation $\cos\phi + i\sin\phi$, for in which case De Moivre's formula $e^{i\phi} \cdot e^{i\psi} = e^{i(\phi+\psi)}$ appears merely as a consequence of the elementary addition theorem of trigonometry.

From this formal point of view, in order to make the relation $e^x \cdot e^y = e^{x+y}$ e*. e* remain valid for any complex arguments, we set up the further definition

$$e^x = e^{\xi}(\cos\eta + i\ \text{xin}\,\eta)$$

where,

$$x = \xi + i\eta\ (\xi, \eta \text{ real}).$$

GENERAL THEORY OF FUNCTIONS OF A COMPLEX VARIABLE

Although the purely formal point of view, indicated above, is by itself free from objection, it is still desirable to recognize in the above formula something more than a mere formal connection. In order to pursue this aim, we are led to the general theory of functions, as (for the sake of brevity) we call the general theory of the so-called analytic functions of a complex variable.

In this, we may use as our starting point a general discussion of the theory of power series with complex variables and complex coefficients. The construction of such a theory of power series presents no difficulties once we have defined the concept of limit in the domain of complex numbers; in fact, it follows almost exactly the theory of real power series.

But as we shall not make any use of these matters in what follows, we shall content ourselves here by stating certain facts and omit the proofs. If a power series converges for any complex value $x = \xi$ whatever, then it converges absolutely for every value x for which $|x| < |\xi|$; if it diverges for a value $x = x$, then it diverges for every value x for which $|x| > |\xi|$.

A power series, which does not converge everywhere, but does converge for some other point in addition to $x = 0$, has a circle of convergence, that is, there exists a number $\rho > 0$ such that the series converges absolutely for $|x| < \rho$ and diverges for $|x| > \rho$.

Having established the concept of functions of a complex variable, represented by power series, and having developed the rules for operating with such functions, we can think of the functions e^x, sin x, cos x, artan x, etc., of the complex variable x as simply defined by the power series which represent them for real values of x.

Then, all the above formal relationships reduce to trivialities. We shall merely indicate by two examples, how this introduction of complex variables helps us to understand the elementary functions. The geometric series for l/(l + $x$ýÿ) ceases to converge when x leaves the interval -1 } x } 1 and so does the series for artan x, although there are no peculiarities in the behaviour of these functions at the ends of the interval of convergence; in fact, they and all their derivatives are continuous for all real values of x.

On the other hand, we can readily understand that the series for l/(l – x) and log (1 – x) cease to converge as x passes through the value 1, since they become infinite there. But this divergence of the series for the inverse tangent and the series,

$$\sum_{\nu=0}^{\infty}(-1)v_x 2_\nu \, for |x|>1$$

becomes clear immediately, if we also consider complex values of x. In fact, we find that when $x = i$, the sum functions become infinite and so cannot be represented by a convergent series. Hence, by our theorem about the circle of convergence, the series must diverge for all values of x such that $|x| > |i|$; in particular, for real values of x, the series diverge outside the interval $-1 < x < 1$. Another example is given by the function $f(x) = e^{-1/x2}$ for $x\ \psi 0$, which, in spite of its apparently regular behaviour, cannot be expanded in a Taylor series.

7

Continuity of Functions

DERIVATIVE

The fact that in the limiting process, which defines the derivative, the difference Δx tends to 0 is sometimes expressed by saying that the quantity Δx becomes infinitely small. This expression indicates that the passage to the limit is regarded as a process during which the quantity Δx is never zero, yet approaches zero as closely as we please.

In Leibnitz's notation, the passage to the limit in the process of differentiation is symbolically expressed by replacing the symbol Δ by the symbol d, so that we can define Leibnitz's symbol for the derivative by the equation

$$\frac{dy}{dx} = \lim_{\Delta x \to 0} \frac{\Delta y}{\Delta x}$$

However, if we wish to obtain a clear grasp of the meaning of the differential calculus, we must beware of regarding the derivative as the quotient of two quantities which are actually infinitely small. The difference quotient $\Delta y/\Delta x$ definitely must be formed with differences Δx which are not equal to 0.

After forming this difference quotient, we must imagine the passage to the limit carried out by means of a transformation or some other device. We have no right to assume that first Δx goes through something like a limiting process and reaches a value which is infinitesimally small but still not 0, so that Δx and Δy are replaced by the infinitely small quantities or infinitesimals dx and dy, and the quotient of these quantities is then formed. Such a conception of the derivative is incompatible with the clarity of ideas demanded in mathematics; in fact, it is altogether meaningless. For a great many simple-minded people, it undoubtedly has a certain charm - the charm of mystery - which is always associated with the word infinite; and in the early days of the differential calculus, even Leibnitz himself was capable of combining these vague mystical ideas with a thoroughly clear understanding of the limiting process.

It is true that this fog, which surrounded the foundations of the new science, did not prevent Leibnitz or his great successors from finding the right path. But this does not release us from the duty of avoiding every such hazy idea in our building-up of the differential and integral calculus. However, the notation of Leibnitz is not merely attractive in itself, but is actually of great flexibility and the greatest use.

The reason is that we can deal in many calculations and formal transformations with the symbols *dy* and *dx* in exactly the same way as if they were ordinary numbers. They enable us to give neater expression to many calculations which can be carried out without their use. In the following pages, we shall see this fact verified over and over again and shall find ourselves justified in making free and repeated use of it, provided we do not lose sight of the symbolical character of the signs *dy* and *dx*.

Leibnitz has also devised for the second and higher derivatives a notation of great suggestiveness and practical utility. He thinks of the second derivative as the limit of the second difference quotient in the following manner. In addition to the variable x, we consider $x_1 = x + h$ and $x_2 = x + 2h$. We then take the second difference quotient as meaning the first difference quotient of the first difference quotient, i.e., the expression

$$\frac{1}{h}\left(\frac{y_2 - y_1}{h} - \frac{y_1 - y}{h}\right) = \frac{1}{h^2}\left(y^2 - 2y_1 + y\right)$$

where $y = f(x)$, $y_1 = f(x_1)$ and $y_2 = f(x_2)$. If we also write $h = \Delta x$ and $y_2 - y_1 = \Delta y_1$, $y_1 - y = \Delta y$, we may appropriately call the expression in the last bracket the difference of the difference of y or the second difference of y and write symbolically *

$$y_1 - 2y_1 + y = \Delta y_1 - \Delta y_1 - \Delta y = \Delta(\Delta y) = \Delta^2 y$$

Here $\Delta\Delta = \Delta$ is not a square, but merely a symbol for difference of difference or second difference. In this symbolic notation, the second difference quotient is then $\Delta y/(\Delta x)$, where the denominator is really the square of Δx, while in the numerator the number 2 syrnbolically denotes the repetition of the difference process. This symbolism for the difference quotient led Leibnitz to introduce the notation

$$y'' = f''(x) = \frac{d^2 y}{dx^{2'}} y''' = f'''(x) = \frac{d^3 y}{dx^3}, \&c.$$

for the second and higher derivatives, and we shall find that this notation also stands the test of use. We must emphasize that the statement that the second derivative may be represented as the limit of the second difference quotient requires proof. For we previously defined the second derivative not in this way, but as the limit of the first difference quotient of the first derivative. In actual fact, the two definitions are equivalent, provided the second derivative is continuous; however, the proof is not given, as we have no particular need for it here.

DIFFERENTIABILITY AND CONTINUITY OF FUNCTIONS

It is useful to point out that, if we know that a function can be differentiated, we need not give any special proof of its continuity.

If a function is differentiable, then it is necessarily continuous. In fact, if the difference quotient $\frac{f(x+h)-f(x)}{h}$ approaches a definite limit as h tends to zero, the numerator of the fraction, that is, $f(x+h)-f(x)$, must tend to zero with h, and this fact expresses the continuity of the function $f(x)$ at the point x.

However, the converse of this is definitely false; it is not true that every continuous function has a derivative at every point. The simplest example, which disproves this assumption, is the function $f(x) = |x|$, i.e., $f(x) = -x$ for $x \succ 0$ and $f(x) = x$ for $x \succ 0$; its graph. At the point $x = 0$, this function is continuous, but has no derivative.

The limit of $[f(x+h)-f(x)]/h$ is equal to 1 if h tends to 0 through positive values and equal to - 1 if h tends to zero through negative values; if we do not restrict the sign of h, There does not exist a limit.

We say that the function has different right hand and and left-hand derivatives at the point x, where by we mean the limiting values of $[f(x+h)-f(x)]/h$ as h approaches 0 through positive values only and negative values only, respectively. Thus, the differentiability of a function requires not merely that the right-hand and left-hand derivatives exist, but that they are equal. Geometrically, the inequality of the two derivatives means that the curve has a sharp corner.

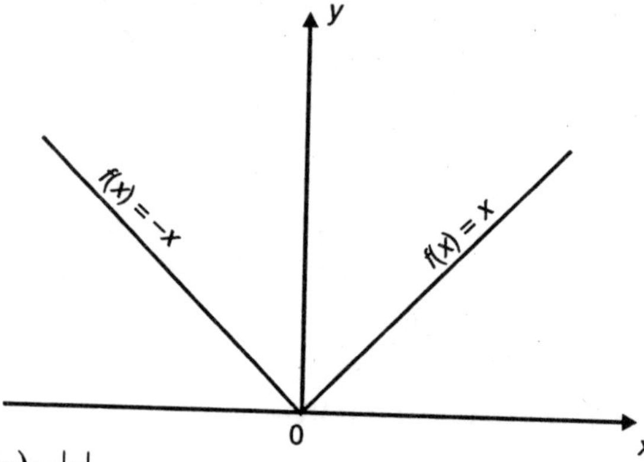

$f(x)=|x|$

As further examples of points where a continuous function is not differentiable, we consider the points where the derivative becomes infinite, i.e., the points at which there exists neither a right-hand nor a left-hand derivative, the difference quotient $[f(x+h)-f(x)]/h$ increasing beyond all bounds as h } 0. For example, the function $y=f(x)=\sqrt[3]{x}=x^{1/3}$ is defined and continuous for all values of x. For all non-zero values of x, its derivative is given by the formula $y'=x^{-2/3}/3$. At the point $x=0$, we have $[f(x+h)-f(x)]/h=h^{1/3}/h=h^{-2/3}$, and we see at once that, as h } 0, the expression has no limiting value, but tends to }.

This state of affairs is often briefly described by saying that the function possesses an infinite derivative or the derivative ∞ at the point in question; however, we should remember that this merely means that, as h tends to 0, the difference quotient increases beyond all bounds and that the derivative in the sense, in which we have defined it, really does not exist.

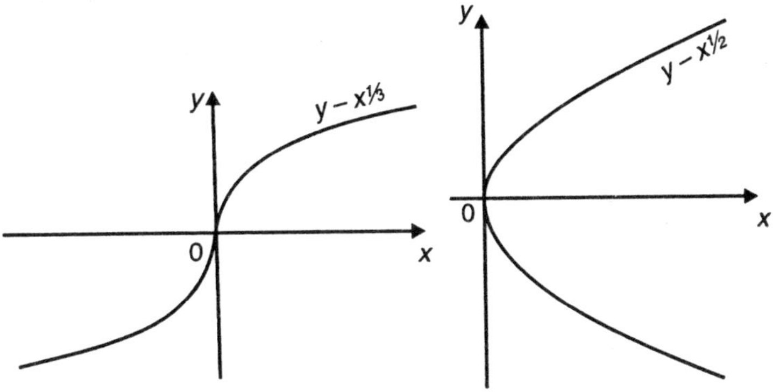

The geometrical meaning of an infinite derivative is that the tangent to the curve is vertical.

The function $y = f(x) = \sqrt{x}$ which is defined and continuous for $x \geq 0$, is also non-differentiable at the point $x = 0$. Since y is undefined for negative values of x, we here consider only the right-hand derivative.

The equation $\dfrac{f(h) - f(0)}{h} = \dfrac{1}{\sqrt{h}}$ shows us that this derivative is infinite; the curve touches the y-axis at the origin.

Finally, the function $y = \sqrt[3]{x^2} = x^{2/3}$ is a case in which the right-hand derivative at the point $x = 0$ is positive and infinite, while the left-hand derivative is negative and infinite, as follows from the relation

$$\frac{f(h) - f(0)}{h} = \frac{1}{\sqrt[3]{h}}$$

As a matter of fact, the continuous curve $y = x^{2/3}$, the so-called semi-cubical parabola or Neil's parabola, has at the origin a cusp, perpendicular to x-axis.

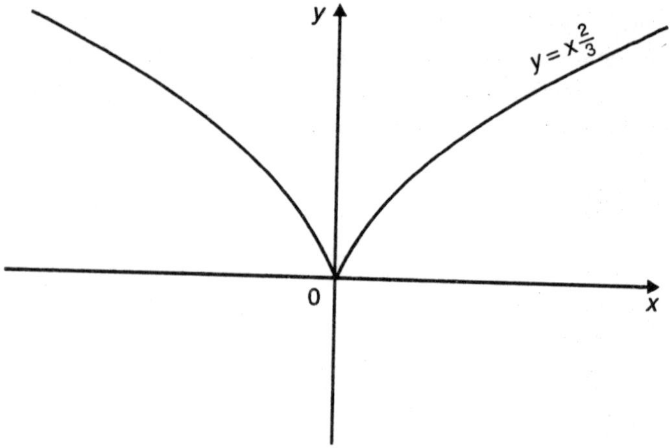

The concept of the derivative, like that of the integral, has an intuitive origin. Its sources are:

1. The problem of construction of the tangent to a given curve at a given point and
2. The problem of finding a precise definition for the velocity during an arbitrary motion.

THE DERIVATIVE AS A VELOCITY

Just as naive intuition led us to the notion of the direction of the tangent to a curve, so it causes us to assign a velocity to a motion. The definition of velocity leads us once again to exactly the same limiting process which we have already called differentiation.

For example, consider the motion of a point along a straight line, the position of the point being determined by a single co-ordinate y, which is the distance, with its proper sign, of our moving point from a fixed point on the line. The motion is given, if we know y as a function of the time t, $y = f(t)$. If this function is a linear function $f(t) = ct + b$, we speak of uniform motion with the velocity c, and for every pair of values t and t_1, which are not equal to each other, we can write

$$c = \frac{f(t_1) - f(t)}{t_1 - t}$$

The velocity is therefore the difference quotient of the function $ct + b$, and this difference quotient is completely independent of the particular pair of instants considered. But what are we to understand by the velocity of motion at an instant t if the motion is no longer uniform?

In order to arrive at this definition, we consider the difference quotient $\frac{f(t_1)-f(t)}{t_1-t}$ which we shall call the average velocity in the time interval between t_1 and t. If now this average velocity tends to a definite limit when we let the instant t_1 come closer and closer to t, we shall naturally define this limit as the velocity at the time t. In other words: The velocity at the time t is the derivative

$$f'(t) = \lim_{t_1 \to t} \frac{f(t_1)-f(t)}{t_1-t}$$

From this new meaning of the derivative, which in itself has nothing to do with the tangent problem, we see that it really is appropriate to define the limiting process of differentiation as a purely analytical operation independently of geometrical intuitions. Here again, the differentiability of the position-function is an assumption which we shall always make tacitly and which, in fact, is absolutely necessary if the notion of velocity is to have any meaning. As a simple example of the connection between motion and velocity, consider the case of a freely falling body. We begin with the experimentally established law that the distance traversed in time t by a freely falling body is proportional to t and therefore can be represented by a function of the form

$$y = f(t) = at^2$$

As before, we find immediately that the velocity is given by the expression $f'(t) = 2at$, which shows that the velocity of a freely falling body increases proportionally to the time.

Examples: We now proceed to work out a number of examples of the actual differentiation of functions and begin with the function $y = f(x)$, where c is a constant.

It is then always true that $f(x + h) - f(x) = c - c = 0$, so that

$\lim_{t_1 \to 0} \dfrac{f(x+h)-f(x)}{h} = 0$ that is, the derivative of a constant is zero.

For a linear function $y = f(x) = cx + b$, we find that

$$\lim_{t_1 \to 0} \frac{f(x+h)-f(x)}{h} = \lim_{h \to 0} \frac{ch}{h} = c$$

Moreover, we shall differentiate the function

$$y = f(x) = x^{\alpha},$$

at first assuming that α is a positive integer. Provided x_1 x, we have

$$\frac{f(x_1)-f(x)}{x_1 - x} = \frac{x_1^{\alpha} - x^{\alpha}}{x_1 - x};$$

the right-hand side of this equation is equal to $x_1^{\alpha-1} + x_1^{\alpha-2}x + \ldots + x^{\alpha-1}$, as we see either by direct division or by using the formula for the sum of a geometric progression.

The new expression for the right-hand side of the equation is a continuous function, whence we can carry out the passage to the limit (x_1 } x) by simply replacing x_1 everywhere by x. Each term is then $x^{\alpha-1}$ and since the number of terms is exactly α, we obtain

$$y' = f'(x) = \frac{\alpha\left(x^{\alpha}\right)}{dx} = \alpha x^{\alpha-1}$$

We arrive at the same result if α is a negative integer $-\beta$; however, we must assume that x is not zero. We then find that

$$\frac{f(x_1)-f(x)}{x_1 - x} = \frac{\frac{1}{x_1^{\beta}} - \frac{1}{x^{\beta}}}{x_1 - x} = -\frac{x^{\beta} - x_1^{\beta}}{x - x_1} \cdot \frac{1}{x^{\beta} x_1^{\beta}}$$

$$= -\frac{x^{\beta-1} + x^{3-2} x_1 + \ldots + x_1^{\beta-1}}{x_1^{\beta} x^{\beta}}$$

Once again, we can carry out the passage to the limit simply by replacing x_1] everywhere by x. Then, just as above, we obtain for the limit the expression

$$y' = -\beta \frac{x^{\beta-1}}{x^{2\beta}} = -\beta x^{-\beta-1}$$

Hence, for negative integral values of α, the derivative is again given by

$$y' = \alpha x^{\alpha-1}$$

Finally, we shall prove the same formula, where x is positive and α any rational number. We let $\alpha = p/q$, where both p and q are integers and, moreover, positive. (If one of them were negative, no essential changes in the proof would be required; for $\alpha = 0$, the result is already known, since x^{α} is then constant.)

We now have,

$$\frac{f(x_1)-f(x)}{x_1-x} = \frac{x_1^{p/q} - x^{p/q}}{x_1 - x}$$

If we now let $x^{1/q} = \xi$ and $x_1^{1/q} = \xi_1$, we obtain

$$\frac{f(x_1)-f(x)}{x_1-x} = \frac{\xi_1^p - \xi^p}{\xi_1^q - \xi^q} = \frac{\xi_1^{p-1} + \xi_1^{p-2}\xi + \ldots + \xi^{p-1}}{\xi_1^{q-1} + \xi_1^{q-2}\xi + \ldots + \xi^{q-1}}$$

After this last transformation, we can immediately perform the passage to the limit ($x_1 \to x$ or, what amounts to the same thing. $\xi_1 \to \xi$), and thus obtain for the limiting value

$$y = \frac{p}{q}\frac{\xi_1^{p-1}}{\xi_1^{q-1}} = \frac{p}{q}\xi^{p-q} = \frac{p}{q}x^{(p-q)/q} = \frac{p}{q}x^{p/q-1}$$

which is formally the same result as before. We leave it to the reader to prove that the same differentiation formula holds also for negative rational superscripts. Once we have developed the theory in a more connected form.

Finally, as a further example, we consider the differentiation of the trigonometric functions: sin x and cos x. We use the elementary trigonometrical formula

$$\frac{\sin(x+h)-\sin x}{h} = \frac{\sin x \cosh + \cos x \sinh - \sin x}{h}$$

$$= \sin x \frac{\cosh - 1}{h} + \cos x \frac{\sinh}{h}$$

we know that

$$\lim_{h\to 0} \frac{\sinh}{h} = 1, \ \lim_{h\to 0} \frac{\cosh - 1}{h} = 0$$

Thus, we find for the required derivative immediately,

$$y' = \frac{d(\sin x)}{dx} = \cos x$$

The function $y = \cos x$ can be differentiated in exactly the same manner. Starting with,

$$\frac{\cos(x+h) - \cos x}{h} = \cos x \frac{\cosh - 1}{h} - \sin x \frac{\sinh}{h}$$

and taking the limit as $h \rightarrow 0$, we at once obtain the derivative,

$$y' = \frac{d(\cos x)}{dx} = -\sin x$$

THE DERIVATIVE AND THE TANGENT

We shall first deal with the tangent problem. If P is a point on a given curve, we shall, in conformity with naive intuition, define the tangent to the curve at the point P by means of the following geometrical limiting process. In addition to the point P, we consider a second point P_1 on the curve. Through the two points P and P_1, we draw a straight line, a secant of the curve. If we now let the point P_1 move along the curve towards the point P, this secant will tend to a limiting position which is independent of the direction from which it approaches P.

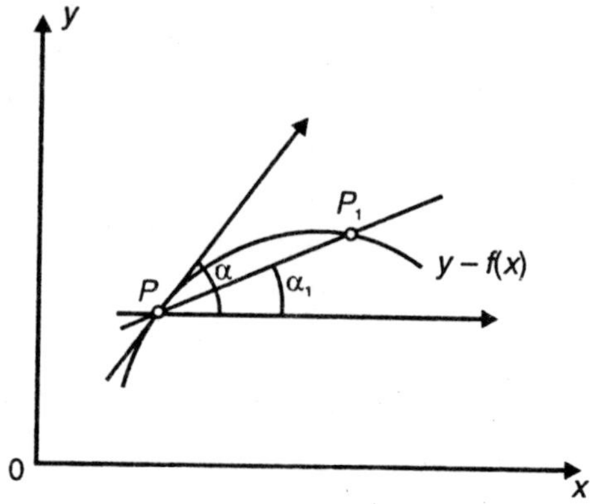

Fig. Chord and Tangent

This limiting position of the secant is the tangent and the

statement that such a limiting position of the secant exists is equivalent to the assumption that the curve has a definite tangent or a definite direction at the point P.

Once we have represented our curve by means of a function $y = f(x)$, there arises the problem of representing our geometrical limiting process analytically, using the function $f(x)$. We take the angle, which a straight line l makes with the x-axis, as being the angle through which the positive x-axis must be turned in the positive direction in order to be for the first time parallel to the line l. Let α_1 be the angle which the secant PP_1 forms with the positive x-axis and α the angle which the tangent forms with the positive x-axis. Then, disregarding the case of a perpendicular tangent, we obviously have

$$\lim_{P_1 \to P} \alpha_1 = \alpha$$

where the meaning of the symbols is perfectly clear. If $x, y\ (=f(x))$ and $x_1, y_1\ (=f(x_1))$ are the coordinates of the points P and P_1, respectively, we immediately find,

$$\tan \alpha_1 = \frac{y_1 - y}{x_1 - x} = \frac{f(x_1) - f(x)}{x_1 - x}$$

and thus our limiting process is represented by the equation

$$\lim_{x_1 \to x} \frac{f(x_1) - f(x)}{x_1 - x} = \tan \alpha$$

That is, in such a direction that a rotation of $\pi/2$ brings it into coincidence with the positive y-axis, in other words, counter-clockwise.

In order that this equation may have a meaning, we must assume that $0 < |x - x_1| < \delta$, δ being sufficiently small. In what follows, corresponding assumptions will often be made tacitly in the steps leading to limiting processes.

THE EXPRESSION

$$\frac{f(x_1) - f(x)}{x_1 - x} = \frac{y_1 - y}{x_1 - x} = \frac{\Delta y}{\Delta x}$$

is called the difference quotient of the function $y = f(x)$, since the symbols Δy and Δx denote the differences of the function $y = f(x)$ and of the independent variable x, respectively. (the symbol Δ is an abbreviation for the difference and is not a factor!) The tangent of α, the direction angle of the curve,* is therefore equal to the limit to which the difference quotient of our function tends when x_1 tends to x.

The slope or gradient of the curve is given by tan α, whence also the term gradient is used for the derivative of the function represented by the curve.

We call this limit the derivative or differential coefficient of the function $y = f(x)$ at the point x and, as Lagrange did, denote it by the symbol $y' = f'(x)$ or, as Leibnitz did, the symbol dy/dx or $df(x)/dx$ or $d/dx\, f(x)$. We shall discuss the meaning of Leibnitz's notation in greater detail; here we just point out that the notation $f'(x)$ expresses the fact that the derivative is itself a function of x, since it has a definite value for each value of x in the interval under consideration.

This fact is sometimes emphasized by the use of the terms derived function, derived curve. We again quote the definition of the derivative:

$$f'(x) = \lim_{x_1 \to x} \frac{f(x_1) - f(x)}{x_1 - x},$$

or

$$\frac{dy}{dx} = \frac{df(x)}{dx} =$$

$$f'(x) = \lim_{x_1 \to x} \frac{f(x_1) - f(x)}{x_1 - x} = \lim_{\Delta x \to 0} \frac{\Delta y}{\Delta x}$$

$$= \lim_{h \to 0} \frac{f(x+h) - f(x)}{h},$$

where we have replaced in the last expression x_1 by $x + h$. It is impossible to find the derivative merely by putting $x_1 = x$ in the expression for the difference quotient, for then the numerator and denominator would both be equal to 0 and we should then led to the meaningless expression 0/0. On the contrary, the actual performance of the passage to the limit in each individual case depends on certain preliminary steps (transformation of the difference quotient).

For example, for the function $f(x) = x$, we have,

$$\frac{f(x_1) - f(x)}{x_1 - x} = \frac{x_1^2 - x^2}{x_1 - x} = x_1 + x.$$

This function $x_1 + x$ is not the same function as $\frac{x_1^2 - x^2}{x_1 - x}$, because the function $x_1 + x$ is defined at one point where the quotient $\frac{x_1^2 - x^2}{x_1 - x}$ is not defined, namely, at the point $x_1 = x$.

For all other values of x_1, the two functions are equal to each other, whence in the above passage to the limit, in which we specifically required that $x_1 \neq x$, we obtain the same value $\lim\limits_{x_1 \to x} \frac{x_1^2 - x^2}{x_1 - x}$ as for $\lim\limits_{x_1 \to x} (x_1 + x)$ However, since the function $x_1 + x$ is defined and continuous at the point $x_1 = x$, we can do with it what we could not do with the quotient, namely, pass to the limit by simply putting $x_1 = x$. We then obtain for the derivative the expression

$$f'(x) = \frac{d(x^2)}{dx} = 2x$$

The performance of such a process, i.e., the actual formation of the derivative is called the differentiation of the function $f(x)$. We shall see later on how this process of differentiation can actually be carried out in all important cases.

Now, the fact that the problem of differentiating a given function has a definite meaning apart from the geometrical intuition of the tangent is of great significance.

The reader will recall that, in the case of the integral, we freed ourselves from the geometrical intuition of area and, on the contrary, based the notion of area on the definition of the integral. Now, independently of the geometrical representation of a function

$y = f(x)$ by means of a curve, we shall define the derivative of the function $y = f(x)$ as being the new function $y' = f'(x)$, given by the equation above, provided always that the limit of the difference quotient exists.

If this limit exists, we say that the function $f(x)$ is differentiable. From now on, we shall assume that every function dealt under consideration is differentiable unless specific mention is made to the contrary. It should be observed that, if the function $f(x)$ is to be differentiable at the point x, the limit as $h \to 0$ of the quotient $[f(x + h) - f(x)]/h$ must exist independently of the manner in which h tends to 0, whether through positive or through negative values or without restriction as to the sign.

Once we have found the derivative $f'(x)$, we take the direction which makes an angle α with the positive x-axis given by the equation $\tan \alpha = f'(x)$ as the direction of the tangent to the curve at the point (x, y). We thus avoid the difficulties which arise out of the indefiniteness of the geometrical view, since we base the geometrical definition on the analytical one and not *vice versa*.

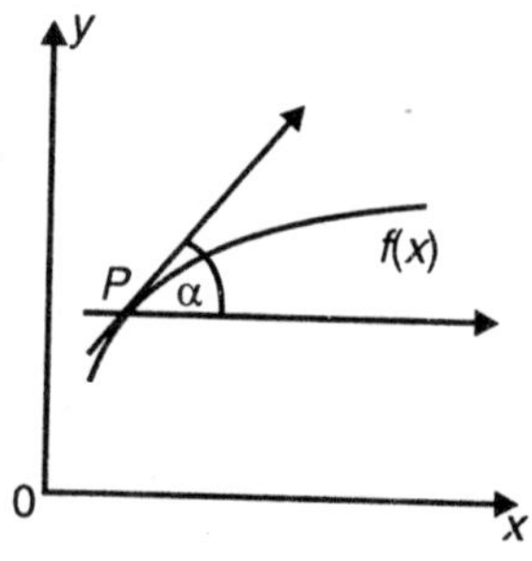

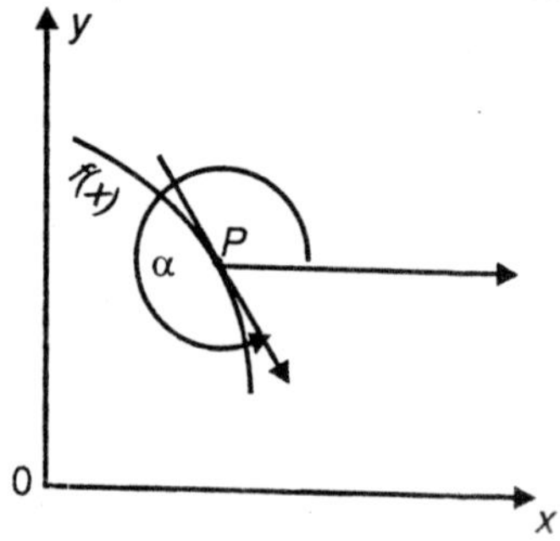

Nevertheless, the visualization of the derivative as the tangent to the curve is an important aid to understanding, even in purely analytical discussions. Accordingly, we shall at once accept the statement based on geometrical intuition:

If $f'(x)$ is positive and the curve is traversed in the direction of increasing x, then the tangent slants upwards, and therefore at the point in question the curve rises as x increases; on the other hand, if $f'(x)$ is negative, the tangent slants downwards and the curve falls as x increases.

Analytically, this follows from the remark that the limit of $[f(x+h)-f(x)]/h$ cannot be positive unless the function is increasing at the point x, by which we mean that for all values of h sufficiently close to 0 the value of $f(x+h)$ is greater or smaller than $f(x)$ according to whether h is positive or negative. We can, of course, make a corresponding statement for the case when $f'(x)$ is negative.

HIGHER DERIVATIVES AND THEIR SIGNIFICANCE

The derivative $f'(x)$ of a function is itself a function of x, the graph of which we call the derived curve of the given curve. For example, the derived curve of the parabola $y = x$ is a straight line, represented by the function y = 2x.

The derivede curve of $y = \sin x$ is $y = \cos x$; similarly, the derived curve of $y = \cos x$ is $y = -\sin x$. (Any of these latter curves can be obtained from the others by translation in the direction of the x-axis. It is now quite a natural step to form the derived curves of the derived curves, i.e., to form the derivative of the function $f'(x) = \phi(x)$. This derivative

$$\phi'(x) = \lim_{h\to 0} \frac{f'(x+h) - f'(x)}{h}$$

provided that it really exists, we shall call the second derivative of the function $f(x)$ and denote it by $f''(x)$. Similarly, we may attempt to form the derivative of $f''(x)$, the so-called third derivative of $f(x)$, which we then denote by $f'''(x)$. In the case of most functions of importance, there is nothing to stop us from imagining this process repeated as many times as we like and from thus defining the n-th derivative $f^{(n)}(x)$. At times, it will be convenient to call the function $f(x)$ its own zero-th derivative.

The terms second, third, n-th differential coefficient are also employed. If the independent variable is interpreted as the time t and the motion of a point is represented by means of the function $f(t)$, the physical meaning of the second derivative is found to be the velocity with which the velocity $f'(t)$ changes, or, as it is usually called, the acceleration. Later on, we shall discuss in detail the geometrical interpretation of the second derivative. However, we may note here the following facts: At a point where $f''(x)$ is positive, $f'(x)$ increases with x; on the other hand, if $f''(x)$ is negative, $f'(x)$ decreases with x.

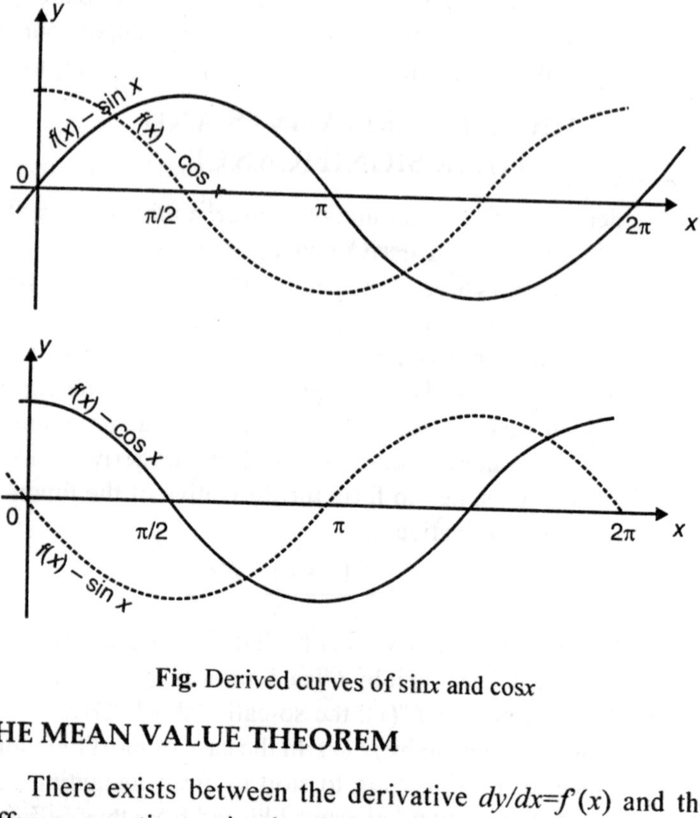

Fig. Derived curves of sin*x* and cos*x*

THE MEAN VALUE THEOREM

There exists between the derivative $dy/dx = f'(x)$ and the difference quotient a simple relation which is important for many purposes is known as the mean value theorem and is obtained in the following manner. We consider the difference quotient

$$\frac{f(x_1) - f(x_2)}{x_1 - x_2} = \frac{\Delta f}{\Delta x}$$

of a function $f(x)$ and assume that the derivative exists everywhere in the interval $x_1 < x < x_2$ so that the graph of the curve has everywhere a tangent. The difference quotient will be represented by the direction of the secant in fact, it is the tangent of the angle α.

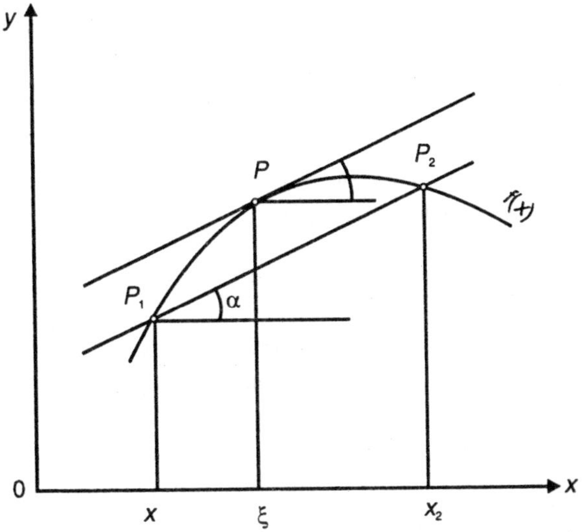

Fig. To illustrate the mean value theorem

Let us imagine that this secant is shifted parallel to itself. At least once, it will reach a position in which it is a tangent to the curve at a point between x_1 and x_2, namely at the point of the curve which is at the greatest distance from the secant. Hence there will be an intermediate value ξ such that

$$\frac{f(x_1)-f(x_2)}{x_1-x_2} = f'(\xi)$$

This statement is called the mean value theorem of the differential calculus. We can also express it somewhat differently by noticing that the number ξ may be written in the form

$$\xi = x_1+\theta(x_2-x_1)$$

where θ is a certain number between 0 and 1. In applications of the mean value theorem, we shall often find that θ cannot be determined more accurately than this, but it will usually turn out that a more accurate value is not required. When accurately formulated, the mean value theorem is:

If $f(x)$ is continuous in the closed interval $x_1 \leq x \leq x_2$ and differentiable at every point of the open interval $x_1 < s < x_2$, then there is at least one value θ, where $0 < \theta < 1$ such that

$$\frac{f(x_2)-f(x_1)}{x_2-x_1} = f'\{x_1+\theta(x_2-x_1)\}$$

If we replace x_1 by x and x_2 by $x + h$, we can express the mean value theorem by the formula

$$\frac{f(x+h)-f(x)}{h} = f'(\xi)=f'(x+\theta h), x<\xi<x+h$$

We wish to emphasize that, while it is essential that $f(x)$ should be continuous for all points of the interval, including the end-points, we need not assume that the derivative exists at the end-points. This apparently trivial remark is actually useful in many applications.

If at any point in the interior of an interval the derivative fails to exist, the mean value theorem is not necessarily true. This is shown by the example $f(x)=|x|$. We can complete our intuitive argument by the following consideration. There is at least one point P on the curve which has the greatest possible distance from the chord joining the points on the curve with the abcissae x_1 and x_2.

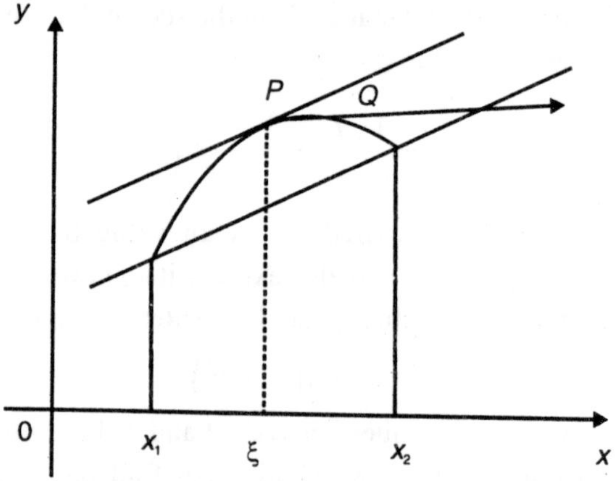

Fig. To illustrate the mean value theorem

At this point, the curve has by assumption a definite tangent. We shall now prove that this tangent must be parallel to the chord. By definition, the tangent is the limiting position of the secant and is obtained by joining P to a point Q on the curve and letting the point Q move towards P. Since, by assumption, Q is not further from the chord than P, the line PQ produced in the direction P to Q must either intersect the chord or run parallel to it; and this must be the case, no matter on which side of P lies the point Q.

This, however, is only possible if the limiting position is parallel to the chord. If we denote the abscissa of the point P by ξ, the slope $f'(\xi)$ of the tangent at P is then equal to the slope of the chord, $[f(x_1)-f(x_2)]/[x_1-x_2)]$, whence we may simply take for the number ξ in the theorem the abscissa of P.

The rigorous proof of the mean value theorem is usually developed as follows: We first establish Rolle's theorem - a special case of the mean value theorem: If a function $\phi(x)$ is continuous in the closed interval $x_1 \leqq x \leqq x_2$ and differentiable in the open interval $x_1 < x < x_2$, and, moreover, $\phi(x_1)=0$ and $\phi(x_1)=0$, then there exists at least one point ξ in the interior of the interval at which $f'(\xi)=0$.

In fact, there must be at least one point ξ, interior to the interval, at which the function $\phi(x)$ takes on its greatest or its least value; to be specific, we assume that ξ is a point where $\phi(x)$ is a maximum so that for every x in the interval $\phi(x) \leqq \phi(\xi)$. Then it is certainly true for every number, the absolute value $|h|$ of which is small enough, that $\phi(\xi)-\phi(\xi+h) \geqq 0$ If h is positive,

$$\frac{\phi(\xi+h)-\phi(\xi)}{h} \leqq 0$$

we now let h tend to zero through positive values and obtain $\phi'(\xi) \leqq 0$. On the other hand, if h is negative,

$$\frac{\phi(\xi+h)-\phi(\xi)}{h} \geqq 0$$

and thus, by letting h tend to zero through negative values, we obtain $\phi'(\xi) \geqq 0$; comparing this result with the preceding inequality, we see that $\phi'(\xi)=0$, which establishes our theorem.

We now apply Rolle's theorem to the function,

$$\phi(x) = f(x)-f(x_1)-\frac{x-x_1}{x_2-x_1}\{f(x_2)-f(x_1)\}$$

which, apart from a factor independent of x, is the distance of the point $(x, f(x))$ of the curve from the secant, as the reader will readily verify. Obviously, this function satisfies the condition $\phi(x) = f(x) + ax + b = 0$ with constant coefficients $a = -[f(x_2) - f(x_1)]/[x_2 - x_1]$ and b. We know already that

$$\phi'(x) = f'(x) + a$$

whence, by Rolle's theorem,

$$0 = \phi'(\xi) = f'(\xi) + a$$

for a suitably chosen intermediate value ξ, and therefore

$$f'(\xi) = -\alpha = \frac{f(x_2) - f(x_1)}{x_2 - x_1}$$

and the mean value theorem has been proved.

As the first of many applications of the mean value theorem,we shall prove the following: Let the function $f(x)$ be continuous in the closed interval $a \leq x \leq b$ and have the derivative $f'(x)$ at every point of the open interval $a < x < b$. Then, if $f'(x)$ is positive everywhere in $a < x < b$, the function $f(x)$ is monotonic increasing in the interval $a \leq x \leq b$; and likewise, if $f(x)$ is negative in $a < x < b$, it is monotonic decreasing.

Let $f'(x) > 0$ and $x_1, x_2 > 0$ be any two values of x in the closed interval. Then, by the mean value theorem,

$$f(x_2) - f(x_1) = (x_2 - x_1) f'(\xi)$$

where $x_1 < \xi < x_2$; Since both factors on the right hand side are positive, this proves that $f(x_2) > f(x_1)$, whence $f(x)$ is monotonic increasing. The Approximate Representation of Arbitrary Functions by Linear Functions. Differentiation: The equation

$$\lim_{h \to 0} \frac{f(x+h) - f(x)}{h} = f'(x)$$

which defines the derivative is equivalent to the equations

$$f(x+h) - f(x) = hf'(x) + \in h$$

$$\text{or } y + \Delta y = f(x + \Delta x) = f(x) + f'(x)\Delta x + \in \Delta x$$

where ε is a quantity which tends to zero with $h = \Delta x$. If, for the moment, we think of the point x as being fixed and the increment Δx as being variable, then, by this formula, the increment of the function, that is, the quantity Δy, consists of two terms, namely a part $hf'(x)$, proportional to h and an error which can be made as small as we please relative to h by making h itself small enough.

Thus, the smaller is the interval about the point x under consideration, the more accurately is the value of the function $f(x + h)$ (which is a function of h), represented by its linear part $f(x) + hf'(x)$. This approximate representation of the function $f(x + h)$ by a linear function of h is expressed geometrically by the substitution of its tangent for the curve at the point x.

Here we merely remark in passing that it is possible to use this approximate representation of the increment Dy by the linear expression $hf'(x)$ to construct a logically satisfactory definition of the notion of a differential as this was done, in particular, by Cauchy.

While the idea of the differential as an infinitely small quantity has no meaning and it is accordingly futile to define the derivative as the quotient of two such quantities, we may still try to assign a sense to the equation $f'(x) = dy/dx$ in such a way that the expression dy/dx need not be thought of as purely symbolic, but as the actual quotient of two quantities dy and dx.

For this purpose, we first define the derivative $f'(x)$ by our limiting process, then think of x as fixed and consider the increment $h = \Delta x$ as the independent variable. We will call this quantity h the differential of x and write $h = dx$.

We now define the expression $dy = y'dx$ as the differential of the function y; dy is therefore a number which has nothing to do with infinitely small quantities.

Thus, the derivative $y' = f'(x)$ is now really the quotient of the differentials dy and dx; however, there is nothing remarkable in this statement; in fact, it is merely a restatement of the verbal definition. The differential dy is accordingly the linear part of the increment Δy.

We shall not make any immediate use of these differentials. Nevertheless, it may be pointed out, for the sake of completeness, that we may also form second and higher differentials. In fact, if we think of h as chosen in any manner, but always the same for every value of x, then $dy = hf'(x)$ is a function of x, of which we can again form the differential.

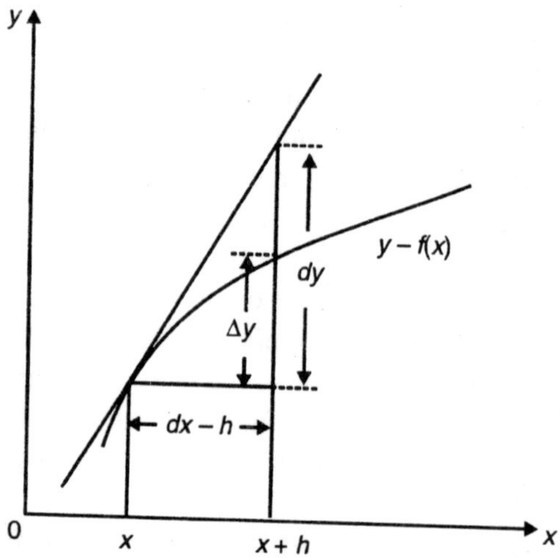

Fig. The Differential dy

The result will be called the second differential of y and be denoted by the symbol $dy = df(x)$. The increment of $hf'(x)$ being $h\{f'(x+h)-f'(x)\}$, the second differential is obtained by replacing the quantity in braces by its linear part $hf''(x)$, so that $dy = hf''(x)$. We may naturally proceed further along the same lines, obtaining third, fourth, differentials of y, etc., which can be defined by the expressions $h^3f'''(x)$, $h^4f^{4)}(x)$, etc.

8

Improper Integrals

FUNCTIONS WITH INFINITE DISCONTINUITIES

It is quite a different matter when a function has an infinite discontinuity inside the interval or at one of its ends. In order to gain even a notion of an integral in this case, we must introduce a further limiting process. Before stating the general definition, we shall illustrate some of the possibilities by means of examples.

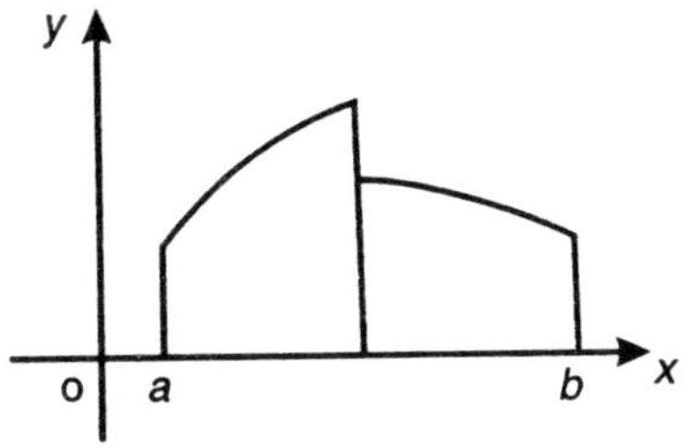

Fig. The Integral of a Discontinuous Function

$$\int \frac{dx}{x^{\alpha}}$$

where α is a positive number. The integrand $1/x^{\alpha}$ becomes infinite as $x \to 0$, whence we cannot extend the integral to the lower limit 0. However, we can try to find what happens as we take the integral from the positive limit ε to the limit 1, say, and finally let $\varepsilon \to 0$. According to the elementary rules of integration, we obtain, provided $\alpha \neq 1$,

$$\int_{\varepsilon}^{1} \frac{dx}{x^{\alpha}} = \frac{1}{1-\alpha}\left(1-\varepsilon^{1-\alpha}\right)$$

We recognize immediately that the following possibilities arise:

- $\alpha > 1$, when, as ε } 0, the right hand side tends to };
- $\alpha < 1$, when the right hand side tends to the limit $1/(1-\alpha)$. hence, in the second case, we shall simply take this limiting value as the integral between the limits 0 and 1. In the first case, we shall say that the integral from 0 to 1 does not exist.
- In the third case, where $\alpha = 1$, the integral will be equal to $-\log \varepsilon$ and therefore, as ε } 0, it approaches no limit, but tends to }, that is, the integral from 0 to 1 does not exist.

Another example of the extension of the integral of a function up to an infinite discontinuity is given by the integrand $\frac{1}{\sqrt{(1-x^2)}}$. We find that

$$\int_0^{1-\varepsilon} \frac{dx}{\sqrt{(1-x^2)}} = \arcsin(1-\varepsilon)$$

If we let ε tend to 0, the right hand side converges to the definite limit $\pi/2$; we therefore call this the value of the integral $\int_0^1 \frac{dx}{\sqrt{(1-x^2)}}$, even though the integrand becomes infinite at the point $x = 1$.

In order to extract a perfectly general concept from these examples, we note, in the first place, that it clearly makes no essential difference whether the discontinuity of the integrand lies at the upper or the lower end of the interval of integration. We now make the following statement: If in an interval a } x } b the function $f(x)$ is continuous with the single exception of the end-point b, we define $\int_a^b f(x)dx$ as the limit

$$\lim_{\varepsilon \to 0} \int_a^{b-\varepsilon} f(x)dx$$

when the point $b - \varepsilon$ approaches the end-point b from inside the interval provided such a limit exists. In this case, we say that the improper integral $\int_a^b f(x)dx$ converges.

However, if no such limit exists, we say that the integral $\int_a^b f(x)dx$ does not exist or does not converge or that it diverges.

An analogous definition holds for the case where the lower limit of the interval of integration instead of the upper one is the exceptional point. Even improper integrals can be intepreted as areas. In the first instance, of course, there is no sense in speaking of the area of a region which estends to infinity; yet, one may attempt to define such an area by means of a passage to the limit from a bounded region with a finite area. For example, the above results for the function $1/x^{\alpha}$ imply that the area, bounded by x axis, the line $x=\varepsilon$ and the curve $y = 1/x^{\alpha}$ tends to a finite limit as ε 0, provided that $\alpha < 1$, and that it tends to infinity if $\alpha < 1$.

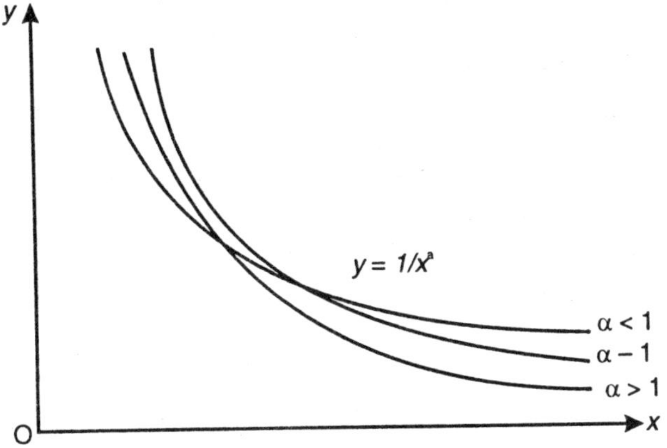

This fact may be simply expressed as follows: The area between the x-axis, the y-axis, the curve and the line $x = 1$ is finite or infinite according to whether $\alpha < 1$ or $\alpha < 1$.

Naturally, intuition can give us no precise information about the finitenees or infiniteness of the area of a region stretching to infinity. We can only say about such a region that the more closely its sides approach one another, the more likely it is to have a finite area.

In this sense, Figure.illustrates the fact that for $\alpha < 1$, the area under our curve remains finite, while it is infinite for $\alpha < 1$. In order to discover whether a function $f(x)$, which has an infinite discontinuity at the point $x = b$, can be integrated up to b, we can often save ourselves a special investigation by means of the criterion:

Let the function $f(x)$ be positive; we will show that this restriction of sign can be removed) in the interval $a < x < b$ and let $\lim_{x \to b} f(x) = \infty$ Then the integral $\int_a^b f(x)dx$ converges, if there exist both a positive number μ less than 1 and a fixed number M independent of x such that everywhere in the interval $a < x < b$ the inequality $f(x) = M/(b - x)^\mu$ is true, in other words, if at the point $x = b$, the function $f(x)$ becomes infinite at a lower order at least than the first.

On the other hand, the integral diverges, if there exist both a number $\nu < 1$ and a fixed number N such that everywhere in the interval $a < x < b$ the inequality $f(x) = N/(b - x)^\nu$ is true, in other words, if at the point $x = b$ the function $f(x)$ becomes infinite to the first order at least. The proof follows almost immediately from a comparison with the very simple special case discussed above. In order to prove the first part of the theorem, we observe that for $0 < \varepsilon < b - a$, we have

$$0 \leqq \int_a^{b-c} f(x)dx \leqq \int_a^{b-c} \frac{M}{(b-x)^\mu} dx$$

As $\varepsilon < 0$, the integral on the right hand side, which is obtained from the integral $\int \frac{dx}{x^\alpha}$, by a simple change of notation, has a limit and therefore remains bounded. Moreover, the values of $\int_a^{b-c} f(x)dx$ increase monotonically as $\varepsilon < 0$; since they are also bounded, they must possess a limit and therefore the integral $\int_a^b f(x)dx$ converges.

The parallel proof of the second part of the theorem is left as an exercise for the reader. We likewise see at once that exactly analogous theorems hold when the lower limit of the integral is a point of infinite discontinuity. If a point of infinite discontinuity lies in the interior of the interval of integration, we merely use this point to subdivide the interval into two parts and then apply the above considerations separately to each of these.

As a further example, we consider the elliptic integral

$$\infty$$

We conclud at oncee from the identity $1 - x = (1 - x)(1 + x)$ that, as $x < 1$, the integrand becomes only infinite at order, whence the improper integral exists.

Infinite Interval of Integration

Another important extension of the concept of integral occurs when one of the limits of integration as infinite. In order to make this extension precise, we introduce the notation: If the integral

$$\int_a^A f(x)dx$$

where a is fixed, tends to a definite limit when A increases positively beyond all bounds, we denote the limit by

$$\int_a^\infty f(x)dx$$

and call it the integral of the function $f(x)$ from a to ∞ . Of course, such an integral does not necessarily exist or, as we often will say, converge.

Simple examples of the various possibilities are again given by the function $f(x)=1/x^{\alpha}$:

$$\int_1^A \frac{dx}{x^\alpha} = \frac{1}{1-\alpha}\left(A^{1-\alpha}-1\right)$$

We see here that, if we again exclude the case $\alpha = 1$, the integral to infinity exists for the case $\alpha > 1$, and, in fact,

$$\int_1^\infty \frac{dx}{x^\alpha} = \frac{1}{\alpha-1}$$

In contrast, when $\alpha < 1$, the integral no longer exists. In the case $\alpha = 1$, the integral again clearly fails to exist, since $\log x$ tends to infinity with x. Hence we see that, as regards integration over an infinite interval, the functions $1/x^{\alpha}$ do not behave in the same way as for integration up to the origin.

This statement also is made plausible by a glance at fiigure. In, we see that, the larger is α, the more slowly do the curves draw in towards the x-axis when x is large, so that we can readily assume that the area under consideration tends to a definite limit for sufficiently large values of α.. The following criterion for the existence of an integral with an infinite limit is often useful. We again assume that for sufficiently large valued of x, say, for $x<\alpha$, the integrand has always the same sign, which, without loss of generality, we can choose to be positive.

Then we have the statement: The integral $\int_a^{\infty} f(x)dx$ converges if the function $f(x)$ vanishes at infinity to a higher order than the first, that is, if there is a number $\nu > 1$ such that for all values of x, no matter how large, the relation $0<f(x) \} M/x^{\nu}$ is true, where M is a fixed number independent of x. Again, the integral diverges, if the function remains positive and vanishes at infinity to an order not higher than the first, that is, if there is a fixed number $N > 0$ such that $xf(x)= N$.

The proof of these criteria, which run exactly parallel to the previous argument, is left to the reader. A very simple example is the integral $\int_a^{\infty} \frac{1}{x^2} dx (a > 0)$. The integrand vanishes at infinity to the second order. As a matter of fact, we see at once that the integral does converge, because $\int_a^{A} \frac{1}{x^2} dx = \frac{1}{a} - \frac{1}{A}$ whence

$$\int_a^{\infty} \frac{1}{x^2} dx = \frac{1}{a}$$

Another equally simple example is

$$\int_0^{\infty} \frac{1}{1+x^2} dx = \lim_{A \to \infty} (\arctan A - \arctan 0) = \frac{\pi}{2}$$

The Gamma Function

Another example of particular importance in analysis is offered by the socalled Γ–function

$$\Gamma(n) = \int_0^\infty e^{-x}x^{n-1}dx \; (n > 0)$$

Here too the criterion of convergence is satisfied; for example, if we choose ν=2, we have since the exponential function e^x tends to zero to a higher order than any power $1/x^m$ ($m > 0$). This gamma function, which we can think of as a function of the number n (not necessarily an integer), satisfies a remarkable relation, which we can arrive at in the following way by integration by parts: To begin with, we have

$$\int e^{-x}x^{n-1}dx = -e^{-x}x^{n-1} + (n-1)\int e^{-x}x^{n-2}dx$$

If we take this formula between the limits 0 and A and then let A increase beyond all bounds, we immediately obtain

$$\Gamma(n) = (n-1)\int_0^\infty e^{-\alpha}x^{n-2}dx = (n-1)\Gamma(n-1)$$

and by this recurrence formula, provided μ is an integer and $0 < \mu < n$,

$$\Gamma(n) = (n-1)(n-2)...(n-\mu)\int_0^\infty e^{-\alpha}x^{n-\mu-1}dx$$

In particular, if n is a positive integer, we have

$$\Gamma(n) = (n-1)(n-2)...3.2.1\int_0^\infty e^{-\alpha}dx$$

and, since

$$\int_0^\infty e^{-\alpha}dx = 1$$

it follows finally that $\Gamma(n) = (n-1)(n-2)...2.1 = (n-1)!$.

This expression for a factorial by an integral is of importanoe in many applications. The integrals,

$$\int_0^\infty e^{-x^2}dx, \int_0^\infty x^n e^{-\alpha^2}dx$$

also converge, as we may easily verify by means of our criterion.

The Dirichlet Integral

A convergent integral, important in many applications, the

convergence of which does not follow directly from our criterion and which is a simple case of a type investigated by Dirichlet, is

$$I = \int_0^{\infty} \frac{\sin x}{x} dx$$

This integral is easily seen to be convergent, if the upper limit is finite, because $\sin x/x \to 1$ as $x \to 0$. Its convergence in the infinite interval is due to the periodic change of sign of the integrand, which causes the contributions to the integral from neighbouring intervals of length π almost to cancel each another. In order to utilize this fact, we write the expression

$$D_{AB} = \int_A^B \frac{\sin}{x} dx$$

in the form

$$D_{AB} = \int_A^{A+\pi} \frac{\sin x}{x} dx - \int_B^{B+\pi} \frac{\sin x}{x} dx + \int_{A+\pi}^{B+\pi} \frac{\sin t}{t} dt$$

introduce in the last of the three integrals the new variable $x = t - p$, whence $\sin t = -\sin x$, and obtain

$$D_{AB} = \int_A^{A+\pi} \frac{\sin x}{x} dx - \int_B^{B+\pi} \frac{\sin x}{x} dx - \int_A^B \frac{\sin x}{x+\pi} dx$$

Addition of this expression to the original one for D_{AB} yields

$$2D_{AB} = \int_A^{A+\pi} \frac{\sin x}{x} dx - \int_B^{B+\pi} \frac{\sin x}{x} dx + \pi \int_A^B \frac{\sin x}{x(x+\pi)} dx$$

Hence, if we assume that $B > A > 0$, it follows that

$$|2D_{AB}| < \frac{2\pi}{A} + \pi \int_A^B \frac{dx}{x^2}$$

in fact, while observing that

$$-\frac{1}{x} \leq \frac{\sin x}{x} \leq \frac{1}{x}$$

and

$$-\frac{1}{x^2} \leq \frac{\sin x}{x(x+\pi)} \leq \frac{1}{x^2}$$

for positive values of x. The integral on the right hand side converges, by our criterion and our formula shows that $|D_{AB}| \to 0$ as both A and B tend to infinity. Now,

$$\left|D_{OB} - D_{OA}\right| = \left|D_{AB}\right|$$

and it follows from Cauchy's convergence test that D_{0B} tends to a definite limit as B. In other words, the integral I exists.

Substitution

It is obvioua that all rules for the substitution of new variables, etc. remains valid for convergent improper integrals. As an example, in order to calculate $\int_0^\infty xe^{-x^2}dx$, we introduce the new variable $u = x$ and obtain

$$\int_0^\infty xe^{-x^2}dx = \frac{1}{2}\int_0^\infty e^{-u}du = \lim_{A\to\infty}\frac{1}{2}\left(1-e^{-A}\right) = \frac{1}{2}$$

Other examples of the use of substitution in the investigation of improper integrals are the Fresnel integrals, which occur in the theory of diffraction of light:

$$F_1 = \int_0^\infty \sin\left(x^2\right)dx, F_2 = \int_0^\infty \cos\left(x^2\right)dx$$

The substitution $x = u$ yields

$$F_1 = \frac{1}{2}\int_0^\infty \frac{\sin u}{\sqrt{u}}du, F_2 = \frac{1}{2}\int_0^\infty \frac{\cos u}{\sqrt{u}}du$$

Integrating by parts, we have

$$\int_A^B \frac{\sin u}{\sqrt{u}}du = \frac{\cos A}{\sqrt{A}} - \frac{\cos B}{\sqrt{B}} - \frac{1}{2}\int_A^B \frac{\cos u}{u^{3/2}}du$$

The integral also tends to 0. Hence, by the same argument as for the Dirichlet integral, we see that the integral F_1 converges. The convergence of the integral F_2 is proved in exactly the same way.

These Fresnel integrals show that an improper integral may exist even although the integrand does not tend to zero as x. In fact, an improper integral can exist even when the integrand is unbounded, as is shown by the example

$$\int_0^\infty 2u\cos\left(u^4\right)du$$

When $u^4 = n\pi$, i.e. when $u = \sqrt[4]{n\pi}$, $n = 0, 1, 2, \ldots$, the integrand becomes $2\sqrt[4]{n\pi}\cos n\pi = \pm 2\sqrt[4]{n\pi}$, so that the integrand is unbounded. However, the substitution $u = x$ reduces the integral to

$$\int_0^{\infty} \cos\left(x^2\right) dx$$

which we have just shown to be convergent.

By means of a substitution, an improper integral may often be transformed into a proper one. For example, the transformation $x = \sin u$ yields

$$\int_0^1 \frac{dx}{\sqrt{\left(1-x^2\right)}} = \int_0^{\pi/2} du = \frac{\pi}{2}$$

On the other hand, integrals of continuous functions may be transformed into improper integrals; this oocurs if the transformation $u = \phi(x)$ is such that at the end of the interval of integration the derivative $\phi'(x)$ vanishes, so that dx/du is infinite.

The Second Mean Value Theorem of the Integral Calculus

The method of integration by parts yields us an easy way for proving an important theorem on the estimation of integrals, usually called the second mean value theorem of the integral calculus.

Let the function $\phi(x)$ be monotonic and continuous in the interval $a \leq x \leq b$, the derivative $\phi'(x)$ be continuous and moreover $f(x)$ be an arbitrary function which is continuous in the same interval. Then one has the second mean value theorem of the integral calculus:

There exists a number ξ such that $a \leq \xi \leq b$ for which

$$\int_a^b f(x)\phi(x)dx = \phi(a)\int_a^{\xi} f(x)dx + \phi(b)\int_{\xi}^{b} f(x)dx$$

We prove this by noting first that we can assume that $\phi(b) = 0$; in fact, replacement of $\phi(x)$ by $\phi(x) - \phi(b)$ changes both sides of the equation by the same amount and yields a function which vanishes at $x = b$. Moreover, we can assume that $\phi(a) > 0$; in fact, if $\phi(a) < 0$, we need only replace $\phi(x)$ by $-\phi(x)$, which changes the sign of both sides of the equation. (The case $\phi(a) = 0$ is trivial;

in fact, if both $\phi(a)$ and $\phi(b)$ vanish, $\phi(x)$ must be identically zero and our equation becomes $0 = 0$.) Hence, we must only prove that, if $\phi(x)$ is continuous and monotonic decreasing and $\phi(b) = 0$, then

$$\int_a^b f(x)\phi(x)dx = \phi(a)\int_a^\xi f(x)dx$$

We now set $F(x) = \int_a^x f(x)dx$ and apply the formula for integration by parts to the left hand side of the last equation; we then have

$$\int_a^x f(x)\phi(x)dx = F(x)\phi(x)\Big|_a^b + \int_a^b F(x)\{-\phi'(x)\}dx$$

The integrated part vanishes, since $F(a)$ and $\phi(b)$ are zero. The expression $-\phi'(a)$ is everywhere positive, so that we can apply the first mean value theorem of the integral calculus. We thus find that the integral on the right hand side has the value

$$F(\xi)\int_a^b\{-\phi'(x)\}dx, a \leqq \xi \leqq b$$

However,

$$F(\xi) = \int_a^\xi f(x)dx$$

and

$$\int_a^b\{-\phi'(x)\}dx = \phi(a) - \phi(b) = \phi(a)$$

and our theorem is established.

This theorem can be extended (although we shall not carry out the proof) to more general classes of functions. The theorem remains true for all continuous monotonic functions $\phi(x)$ whether they have derivatives or not. In fact, it is true for any discontinuous monotonic function for which we are in a position to integrate $f(x)\phi(x)$.

NUMERICAL METHODS

Anyone who has to use analysis as an instrument for investigating physical or technical phenomena is faced by the question whether and how the theory can be adapted to yield useful practical methods for actual numerical calculations. Yet, even from the point of view of the theoretical person, who only desires to recognize the link between natural phenomena and not to conquer them, these questions are of no trifling interest. Here we can only discuss some particularly important points which are more or less closely linked to the preceding ideas. We wish to direct special attention to the fundamental fact that the meaning of an approximate calculation is not precise unless it is accompanied by a definite knowledge of the degree of accuracy attained.

Numerical Integration

We have seen that even relatively simple functions cannot be integrated in terms of elementary functions and that it is futile to make this unattainable goal the aim of the integral calculus. On the other hand, the definite integral of a continuous function does exist and this fact raises the problem of finding methods for calculating it numerically. Here we shall discuss the simplest and most obvious of these methods with the aid of geometrical intuition and thenconsider error estimation.

Our objective is to calculate the integral

$$I = \int_a^b f(x)dx,$$

where $a < b$. We imagine the interval of integration to have been subdivided into n equal parts of length $h = (b - a)/n$ and denote the points of subdivision by $x_1=a, x_2=a+h, \ldots, x_n=b$, the values of the function at the points of subdivision by $f_0, f_1, \ldots, f_n$ and, similarly, the values of the function at the midpoints of the intervals by $f_{1/2}, f_{3/2}, \ldots, f_{(2n-1)/2}$. We interpret our integral as an area and cut up the region under the curve into strips of width h in the usual manner. We must now obtain an approximation for each such strip of surface, that is, for the integral

$$I_v = \int_{x_y}^{x_y+h} f(x)dx,$$

The Rectangle Rule: The crudest and most obvious method of approximating the integral I is directly linked to the definition of the integral; we replace the area of the strip I_n by the rectangle of area $f_n h$ and obtain for the integral the approximate expression

$$I \approx h(f_0 + f_1 + \ldots + f_{n-1})$$

From here on, the symbol ≀ means is approximately equal to.

$$x_\nu,\ x_{\nu+1} = x_\nu + h, \ and\ x_{\nu+2} = x_\nu + 2h$$

The Trapezoid and Tangent Formulae: We obtain a closer approximation with no greater trouble if we do not replace the area of the strip I_ν, as above, by a rectangular area, but by the trapezoid of area $(f_\nu+f_{\nu+1})h/2$. For the whole integral, this process yields the approximate expression

$$I \approx h(f_1 + f_2 + ... + f_{n-1}) + \frac{h}{2}(f_0 + f_n)$$

(trapezoid formula), since, when the areas of the trapezoids are added, each value of the function except the first and the last occurs twice.

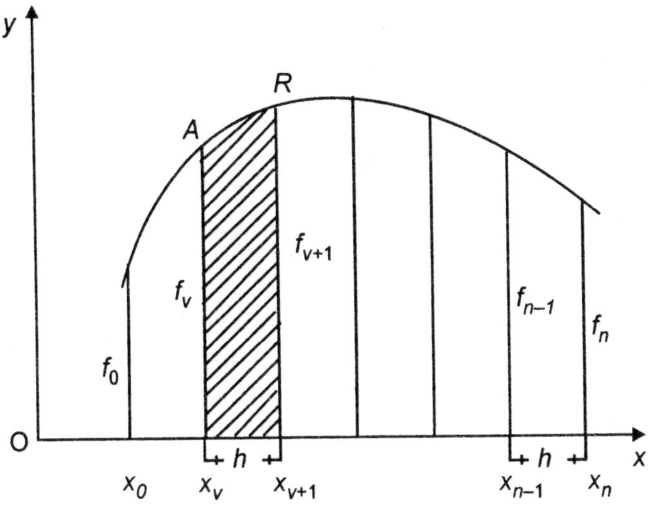

Fig. The Frapezoid Formula

As a rule, the approximation becomes even better if, instead of choosing the trapezoid under the chord *AB* as an approximation to the area of I, we select the trapezoid under the tangent to the curve at the point with the abscissa $x = x_\nu+h/2$. The area of this trapezoid is simply $hf_{\nu+\frac{1}{2}}$, so that the approximation for the entire integral is

$$I \approx h(f_{1/2} + f_{3/2} + ... + f_{2n-1/2}),$$

which is called the tangent formula.

Simpson's Rule: By means of Simpson's rule, we arrive with very little more trouble at a numerical result which is generally much more exact. This rule depends on estimating the area $I_v + I_{v+1}$ of the double strip between the abscissae $x = x_v$ and $x = x_v + 2 = x_{v+2}$ by considering the upper boundary to be no longer a straight line but a parabola - in order to be specific, that parabola which passes through the three points of the curve with the abscissae x_v, $x_{v+1} = x_v + h$, *and* $x_{v+2} = x_v + 2h$. The equation of this parabola is,

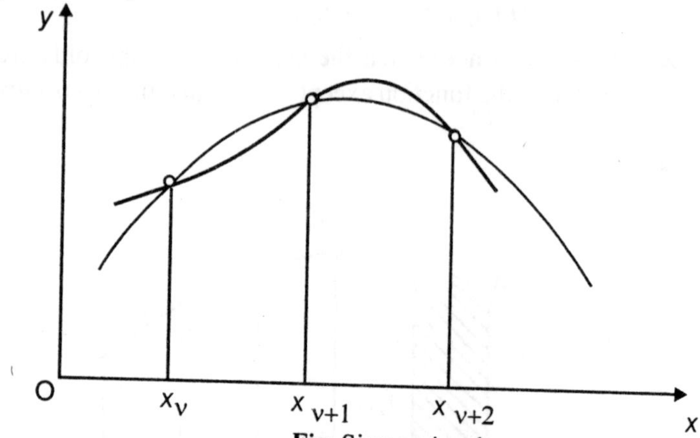

Fig. Simpson's rule

The student should verify by direct substitution that for the three values of x in question this eqnation gives the proper values of y, i.e., f_n, f_{n+1} and f_{n+2}, respectively. If we integrate this second degree polynomial between the x_n and $x_n + 2h$, we obtain, after a brief calculation, for the area under the parabola

$$\int_{xv}^{xv+2h} y\,dx = \begin{array}{l} 2hf_v + 2h(f_{v+1} - f_v) \\ + \frac{1}{2}\left(\frac{8}{3}h - 2h\right)(f_{v+2} - 2f_{v+1} + f_v) \end{array}$$

$$= \frac{h}{3}(f_v + 4f_{v+1} + f_{v+2})$$

This represents the required approximation to the area of our strip $I_n + I_{n+1}$. If we now assume that $n = 2m$, i.e., that n is an even number, we obtain, by addition of the areas of such strips, Simpson's rule,

$$I \approx \frac{4h}{3}(f_1 + f_3 + + \ldots f_{2m-1}) + \frac{2h}{3}(f_2 + f_4 + \ldots + f_{2m-2}) + \frac{h}{3}(f_0 + f_{2m})$$

Estimation of the Error: It is easy to obtain an estimate of the error for each of our methods of integration, if the derivatives of the function $f(x)$ are known throughout the interval of the integration.

We take $M_1, M_2,\ldots$ as the upper bounds of the absolute values of the first, second, derivative, respectively, i.e., we assume that throughout the interval $|f^{(\nu)}(x)| < M_\nu$. Then the estimates are:

For the rectangle rule:

$$\left|I_\nu - hf_\nu\right| < \frac{1}{2}M_1 h^2 or \left|I - h\sum_{\nu=0}^{n-1} f_\nu\right| < \frac{1}{2}M_1 nh^2 = \frac{1}{2}M_1\left(b-a\right)h$$

For the tangent rule:

$$\left|I_\nu - hf_{\nu+\frac{1}{2}}\right| < \frac{M}{24}h^3 \text{ or} \left|I - h\sum_{\nu=0}^{n-1} f_{\nu+\frac{1}{2}}\right| < \frac{M_2}{24}\left(b-a\right)h^2$$

For the trapezoid rule:

$$\left|I_\nu - \frac{h}{2}\left(f_\nu + f_{\nu+1}\right)\right| < \frac{M_2}{12}h^3$$

For Simpson's rule:

$$\left|I_\nu + I_{\nu+1} - \frac{h}{3}\left(f_\nu + 4f_{\nu+1} + f_{\nu+2}\right)\right| < \frac{M_4}{90}h^5$$

The last two estimates also lead to estimates for the entire integral I.

We see that Simpson's rule has an error of much higher order in the small quantity h than the other rules, so that when M_4 is not too large, it is very advantageous for practical calculations. In order to avoid tiring the reader with the details of the proofs of these estimates, which are fundamentally quite simple, we shall restrict ourselves to the proof for the tangent formula. For this purpose, we expand the function $f(x)$ in the $(n + 1)$-th strip by Taylor's theorem:

$$f(x) = f_{\nu+\frac{1}{2}} + \left(x - x_\nu - \frac{h}{2}\right)f'\left(x_\nu + \frac{h}{2}\right) + \frac{1}{2}\left(x - x_\nu - \frac{h}{2}\right)^2 f''(\xi),$$

where ξ is a certain intermediate value in the strip. If we integrate the right hand side over the interval $x_n < x < x_n + h$, the integral of the middle term is zero. Since, as is easily verified,

$$\frac{1}{2}\int_{x_\nu}^{x_\nu+h}\left(x-x_\nu-\frac{h}{2}\right)^2 dx = \frac{h^3}{24},$$

it follows immediately that

$$\left|\int_{x_\nu}^{x_\nu+h} f(x)dx - hf_{\nu+\frac{1}{2}}\right| < M_2\frac{h^3}{24},$$

which proves the above assertion.

Mean Value theorem and of Taylor's Theorem

The Calculus of Errors: We now come to quite a different type of numerical calculations. These are applications of the mean value theorem or, more generally, of Taylor's theorem with remainder or, finally, of the infinite Taylor series.

As an application which, although simple, is quite important in practice, we shall consider the calculus of errors. This rests on the idea - which lies at the root of the entire differential calculus - that a function $f(x)$, which is differentiable a sufficient number of times, can be represented in the neighbourhood of a point by a linear function with an error of order less than the first, by a quadratic function with an error of order less than the second, etc. Consider the linear approximation to a function $y=f(x)$. If $y+\Delta y=f(x+\Delta x)=f(x+h)$, we find by Taylor's theorem

$$\Delta y = \quad hf'(x) + \frac{h^2}{2}f''(\xi)$$

where $\xi = x+\theta h\,(0<\theta<1)$ is an intermediate value, which need not be known more precisely. If $h=\Delta x$ is small, we obtain as a practical approximation $\Delta y \approx hf'(x)$.

In other words, we replace the difference quotient by the derivative to which it is approximately equal and the increment of y by the approximately equal linear expression in h.

We use this fairly obvious fact for practical purposes in the following way. Let two physical quantities x and y be linked by the relation $y=f(x)$.

There arises then the question how an inaccuracy in the

measurement of x affects the determination of y. If we happen to use instead of the true value of x the inaccurate value $x + h$, then the corresponding value of y differs from the true value $y = f(x)$ by $\Delta y = f(x + h) - f(x)$, whence the error is given approximately by the above relation.

Evaluation of π: Gregory's series, also called Leibnitz' series,

$$\frac{\pi}{4} = 1 - \frac{1}{3} + \frac{1}{5} - \frac{1}{7} + - \ldots,$$

using the series for artan, is not suitable for the calculation of π due to the slowness of its convergence. However, we may calculate π with comparative ease by the artifice: From the addition theorem for the tangent

$$\tan(\alpha + \beta) = \frac{\tan\alpha + \tan\beta}{1 - \tan\alpha\ \tan\beta},$$

changing to the inverse functions α = artan u, β = artan v, we obtain

$$\text{arc tan} u + \text{arc tan} v = \text{arc tan}\left(\frac{u+v}{1-uv}\right)$$

If we now choose u and v in such a manner that $(u + v)/(1 - uv) = 1$, we obtain the value $\pi/4$ on the right hand side and, if u and v are small numbers, we can easily compute the left hand side by means of known series. For example, if we set $u = 1/2$, $v = 1/3$, as Euler did, we obtain $\frac{\pi}{4}$ = arc tan $\frac{1}{2}$ +arc tan $\frac{1}{3}$ .Moreover, if we note that $\left(\frac{1}{3} + \frac{1}{7}\right) \div \left(1 - \frac{1}{21}\right) = \frac{1}{2}$ we have arc tan $\frac{1}{2}$ = arc tan $\frac{1}{3}$ +arc tan $\frac{1}{7}$, so that $\frac{\pi}{4}$ = 2 arc tan $\frac{1}{7}$ +arc tan.

Using this formula, Georg Vega (1756 - 1802) calculated the number π to 140 places. Moreover, using the equation,

$$\left(\frac{1}{5} + \frac{1}{8}\right) \div \left(1 - \frac{1}{40}\right) = \frac{1}{3}$$

we obtain arc tan $\frac{1}{3}$ = arc tan $\frac{1}{5}$ +arc tan $\frac{1}{8}$ or $\frac{\pi}{4}$ = 2 arc tan $\frac{1}{5}$ +arc

$\tan\frac{1}{7}+2\text{arc}\tan\frac{1}{8}$. This expansion is extremely useful for the computation of π by means of the series artan$x$ $=x-\frac{x^3}{3}+\frac{x^5}{5}-+\ldots$; in fact, if we substitute for x the values 1/5, 1/7 or 1/8, we obtain with just a few terms a high degree of accuracy, since the terms diminish rapidly. However, we can perform the calculation even more conveniently by basing it on the formula

$$\frac{\pi}{4}=\text{arc}\tan\frac{120}{119}-\text{arc}\tan\frac{1}{239}=4\ \text{arc}\tan\frac{1}{5}-\text{arc}\tan\frac{1}{239}$$

obtained by similar arguments as those above. Calculation of Logarithms: For the computation of logarithms, we transform the series,

$$\frac{1}{2}\log\frac{1+x}{1-x}=x+\frac{x^3}{3}+\frac{x^5}{5}+\ldots(|x|<1);$$

where $0<x<1$, by the substitution

$$\frac{1+x}{1-x}=\frac{p^2}{p^2-1},\ \text{x}=\frac{1}{1p^2-1}$$

into the series

$$\log p=\frac{1}{2}\log(p-1)+\frac{1}{2}\log(p+1)+\frac{1}{2p^2-1}+\frac{1}{3(2p^2-1)^3}+\ldots,$$

where $2p-1>1$, that is, $p>1$. If p is an integer and $p+1$ can be resolved into smaller integral factors, this last series expresses the logarithm of p in terms of the logarithms of smaller integers plus a series, the terms of which diminish very rapidly and the sum of which can therefore be calculated accurately enough from only a few terms. Hence this series allows us to compute the logarithms of any prime number and hence that of any number, provided we have already calculated the value of log 2.

The accuracy of this determination of $\log p$ can be estimated more easily by means of the geometric series than from the general remainder formula.

In fact, we have for the remainder R_n of the series, i.e., the sum of all the terms following the term,

$$\frac{1}{n\left(2p^2-1\right)^n}$$

$$R_n \frac{1}{(n+2)\left(2p^2-1\right)^{n+2}}\left(1+\frac{1}{\left(2p^2-1\right)^2}+\frac{1}{\left(2p^2-1\right)^4}+\ldots\right)$$

$$=\frac{1}{(n+2)\left(2n^2-1\right)^n}\cdot\frac{1}{\left(2p^2-1\right)^2-1}$$

and this formula yields immediately the required error estimate. For example, let us compute $\log_e 7$, using the first four terms of the series. We have

$P = 7, \quad 2P^2 - 1 = 97,$

$\log 7 = 2\log 2 + \frac{1}{2}\log 3 + \frac{1}{97} + \frac{1}{3.97^3} + \ldots, \frac{1}{97} \approx$

$0.01030928, \frac{1}{3.97^3} \approx 0.00000037,$

$2\log 2 \approx 1.38629436, \frac{1}{2}$

$\text{Log}3 \approx 0.54930614,$

whence $\log_e 7 \approx 1.945{,}910{,}15.$

Estimation of the error yields,

$$R_n < \frac{1}{5.94^3} \times \frac{}{97^3-1} < \frac{1}{36\times10^3}$$

However, we must note that each of the four numbers which we have added is only given to within an error, so that the last place in the above value of $\log_e 7$ might be wrong by 2. However, as a matter of fact, the last place is also correct.

Numerical Solution of Equations

In conclusion, we shall add some remarks about the numerical solution of the equation $f(x) = 0$, where $f(x)$ need not necessarily be a polynomial and we are, of course, only concerned with the determination of real roots.

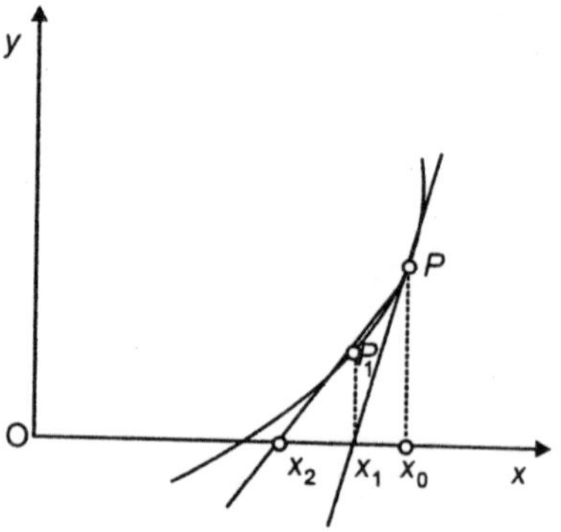

Fig. Newton's Method of Approximation

Every such numerical method is based on the scheme of starting with some known approximation x_0 of one of the roots and then improving this approximation, whence this first approximation for the desired root of the equation is found; the quality of the approximation is not a concern. We may perhaps take a rough guess as a first approximation or, better still, obtain it from the graph of the function $y = f(x)$, the intersection of which with the x-axis yields the required root (of course with an error depending on the scale and the accuracy of the drawing).

Newton's Method: The following procedure, due to Newton, is based on the fundamental principle of the differential calculus - the replacement of a curve by a straight line, the tangent - in the immediate neighbourhood of the point of contact. If we have an approximate value x_0 for a root of the equation $f(x) = 0$, we consider the point on the graph of the function $y = f(x)$ with the co-ordinates $x = x_0$, $y = f(x_0)$. We wish to find the intersection of the curve with the x-axis; we find an approximation at the point $x_0, y_0 = f(x_0)$, where the tangent intersects the x-axis. By virtue of the geometrical meaning of the derivative, Fig. yields at once,

$$\frac{f(x_0)}{x_0 - x_1} = f'(x_0),$$

whence we obtain the formula for the calculation of the new value x_1:

$$x_1 = x_0 - \frac{f(x_0)}{f'(x_0)}$$

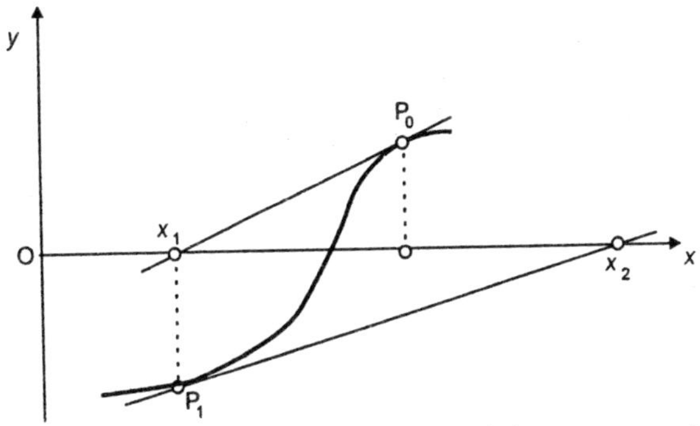

The abscissa x_1 of this intersection of the tangent with the x-axis then represents a new, possibly better approximation to the required root of the equation than x_0. If this procedure yields a better approximation than x_0, we repeat the process and find x_2, etc.; if the curve has the form shown in Fig., these approximations will approach more and more closely to the required solution.

The usefulness of this process depends essentially on the nature of the curve $y=f(x)$. In Fig., we see that the successive estimates converge with greater and greater accuracy to the required root. This is due to the fact that the curve has its convex side turned towards the x-axis. However, we see that if we choose the original value x_0 badly, our construction does not lead at all to the required root. We see from this that while using Newton's method we must examine each individual case in order to determine the degree of accuracy with which we have really solved the equation. We shall return to this topic later on.

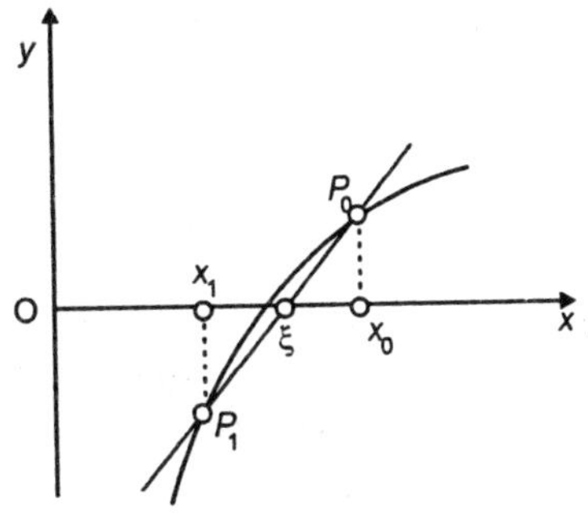

Fig. The Rule of False Position

The Rule of False Position: Newton's method, in which the tangent to the curve has a decisive role, is only the limiting case of an older method, known as the rule of false position, in which the secant takes the place of the tangent. Let us assume that we know two points (x_0, y_0) and (x_1, y_1) in the neighbourhood of the required intersection with the x-axis. If we replace the curve by the secant joining these two points, the intersection of this secant with the x-axis will under certain circumstances be an improved approximation to the required root of the equation. If the abscissa of this point is denoted by ξ, we have the equation

$$\frac{\xi - x_0}{f(x_0)} = \frac{\xi - x_1}{f(x_1)}$$

from which we compute ξ:

$$\xi = \frac{x_0 f(x_1) - x_1 f(x_0)}{f(x_1) - f(x_0)}$$

$$= \frac{x_0 f(x_1) - x_0 f(x_0) + x_0 f(x_0) - x_1 f(x_0)}{f(x_1) - f(x_0)}$$

or

$$\xi = x_0 - \frac{f(x_0)}{\{f(x_1) - f(x_0)\}/(x_1 - x_0)}$$

This formula, which determines the next approximation ξ from x_0 and x_1, is called the rule of false position. We can employ it with advantage if one value of the function is positive and the other negative, where $y_0 > 0$ and $y_1 < 0$. Repetition of this process will always lead us to the required result, if we use at each step a positive and a negative value of the function, between which the required root must necessarily lie.

The above formula of Newton results from the rule of false position as a limiting case, if we let x_1 tend to x_0. In fact, the denominator of the second term on the right hand side of the statement of the rule of false position tends to $f'(x_0)$ as x_1 tends to x_0. The Method of Iteration: Another way of approximating roots of an equation $f(x) = 0$ is the method of iteration. i.e., set $\phi(x) = f(x) + x$ and write it in the form $x = \phi(x)$.

We then assume that ξ is the true value of a solution of our equation and x_0 a first approximation. We obtain a second approximation x_1 by setting $x_1 = \phi(x_0)$, a third approximation x_2 by setting $x_2 = \phi(x_1)$, etc. In order to investigate the convergence of these approximations, we apply the mean value theorem; recalling that $x = \phi(\xi)$, we have

$$\xi - x_1 = \phi(\xi) - \phi(x_0) = (\xi - x_0)\phi'(\xi)$$

where $\bar{\xi}$ lies between ξ and x_0.

This shows that, if for $|\xi - x| < |\xi - x_0|$ the absolute value of the derivative $f'(x)$ is less than $k < 1$,

Then the successive approximations converge, because,

$$|\xi - x_1| < k|\xi - x_0|, |\xi - x_2| < k^2|\xi - x_0|, \ldots, \ |\xi - x_n| <$$

$$k^n|\xi - x_0|,$$

whence the errors tend to zero, The smaller is the absolute value of the derivative $\phi'(x)$ in the neighbourhood of ξ, the more rapid is the convergence. If $\phi'(x) > 1$ in the neighbourhood of ξ, the approximations do not any longer tend to ξ. We can then use the inverse function or else the following device: We choose a first approximation x_0, calculate $A = f'(x_0)$ and write,

$$\phi(x) = -\frac{1}{A}f(x) + x$$

Then the equation $f(x) = 0$ can be written in the form $x = \phi(x)$, and now $\phi'(x) = -f'(x)/A + 1$, which has the value 0 at $x = x_0$, whence its absolute value will usually be less than a constant $k < 1$, provided $|\xi - x| < |\xi - x_0|$. Returning to Newton's method, we can now investigate its suitability for application at any given point. The equation $f(x) = 0$ is equivalent to

$$x = \phi(x) = x - \frac{f(x)}{f'(x)}$$

provided that $f'(x) \neq 0$. Applying the method of iteration to this last equation, we obtain from a first approximation x_0 a second approximation, $x_1 = x_0 - f(x_0)/f'(x_0)$; in other words, the same second approximation as Newton's method gives when it is applied to the equation $f(x) = 0$. We thus see that the smaller is the value of,

$$\phi'(x) = -\frac{f(x)f''(x)}{(f'(x))^2},$$

the more rapidly converge the successive approximations. In words, Newton's formula converges rapidly for large values of $f'(x_0)$ and small values of $f(x_0)$ and curvature, as intuition would lead us to suspect. We can also obtain an estimate of the accuracy of Newton's method, if we recall that, since $f(\xi) = 0$, the derivative $\phi'(\xi) = 0$. Applying Taylor's theorem, we have

$$\xi - x_1 = \phi(\xi) - \phi(x_0) = \frac{(\xi - x_0)^2}{2}\phi''(\xi),$$

where $\bar{\xi}$ lies between ξ and x_0. Thus, if the error of the original estimate is small, the method converges much more rapidly than the method of iteration applied directly to $f(x) = 0$.

For example, if,

$$\phi''(x) = \frac{\{f'(x)\}^2 f''(x) + f'(x)f(x)f'''(x) - 2f(x)\{f''(x)\}^2}{\{f'(x)\}^3} \quad (a)$$

is everywhere less than 10, then a first approximation, which has an error by less than 0.001, will yield a second approximation with an error of less than 0.000,005.

Stirling's Formula

In very many applications, especially in statistics and in the theory of probability, we find it necessary to have a simple approximation to $n!$ as an elementary function of n. Such an expression is given by the following theorem which bears the name of its discoverer James Stirling 1692 - 1770:

As $n \approx \dfrac{n!}{\sqrt{2\pi}n^{n+1/2}e^{-n}} \to 1$; in more detail,

$$\sqrt{2\pi}n^{n+1/2}e^{-n} < n! < \sqrt{2\pi}n^{n+1/2}e^{-n}\left(1 + \frac{1}{4n}\right).$$

In other words, the expressions $n!$ and $\sqrt{2\pi}n^{n+\frac{1}{2}}e^{-n}$ differ only by a small percentage for large n - as we say, the two expressions are asymptotically equal—and at the same time the factor $1 + 1/4n$ gives us an estimate of the degree of accuracy of the approximation.

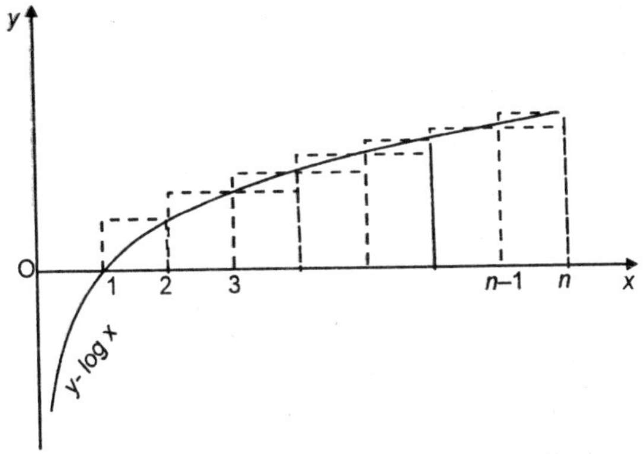

We are led to this remarkable formula, if we attempt to evaluate the area under the curve $y = \log x$. By integration, we find that A_n, the exact area under this curve between the ordinates $x = 1$ and $x = n$, is given by

$$\int_1^n \log x dx = x\log x - x \log x - x\Big|_1^n = n\log n - n + 1$$

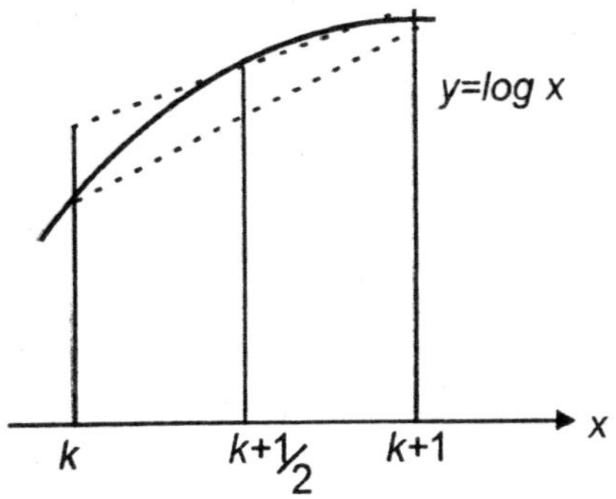

However, if we estimate the area by the trapezoidal rule, erecting ordinates at $x=1, x=2, \ldots, x = n$, we obtain an approximate value T_n for the area:

$$T_n = \log 2 + \log 3 + \ldots + \log(n-1) + \frac{1}{2}\log n = \log n! - \log n$$

If we make the reasonable assumption that A_n and T_n are of the same order of magnitude, we find at once that $n!$ and $n^{n+\cdots}e^{-n}$ are of the same order of magnitude, which is essentially what Stirling's formula states.

In order to make this argument precise, we first show that $a_n = A_n - T_n$ is bounded, whence it follows immediately that $T_n = A_n\left(1-\frac{a_n}{A_n}\right)$ is of the same order of magnitude as A_n.

The difference $a_{k+1} - a_k$ is the difference between the area under the curve and the area under the secant in the strip $k < x < k+1$.

Since the curve is concave downward and lies above the secant, $a_{k+1}-a_k$ is positive, and $(a_n-a_{n-1}) + (a_{n-1}-a_{n-2}) + \dots + (a_2-a_1) + a_1$ is monotonic increasing. Moreover, the difference $a_{k+1}-a_k$ is clearly less than the difference between the area under the tangent at $x = k + 1/2$ and the area under the secant, whence we have the inequality

$$a_{k+1} - ak < \log\left(k+\frac{1}{2}\right) - \frac{1}{2}\log k - \frac{1}{2}\log(k+1)$$

$$= \frac{1}{2}\log\left(1+\frac{1}{2k}\right) - \frac{1}{2}\log\left(1+\frac{1}{\left(k+\frac{1}{2}\right)}\right)$$

$$< \frac{1}{2}\log\left(1+\frac{1}{2k}\right) - \frac{1}{2}\log\left(1+\frac{1}{2(k+1)}\right)$$

If we add these inequalities for $k = 1, 2, \dots, n - 1$, all the terms on the right hand side except two of them will cancel out and (since $a_1 = 0$) we have,

$$a_n < \frac{1}{2}\log\frac{3}{2} - \frac{1}{2}\log\left(1+\frac{1}{2n}\right) < \frac{1}{2}\log\frac{3}{2}$$

Hence a_n is bounded; since it is monotonic increasing, it tends to a limit a as $n$ýý. Our inequality for $a_{k+1} - a_k$ now yields

$$a - a_n = \sum_{k=n}^{\infty}(a_{k+1} - a_k) < \frac{1}{2}\log\left(1+\frac{1}{2n}\right)$$

Since, by definition, $A_n - T_n = a_n$, we now have,

$$\log n! = 1 - a_n + \left(n+\frac{1}{2}\right)\log n - n,$$

or, writing $a_n = e^{1-a}$, $n! = a_n n^{n+\frac{1}{2}} e^{-n}$ The sequence a_n is monotonic decreasing and tends to the limit $a = e^{1-a}$, whence

$$1 < \frac{a_n}{a} = e^{a-a_n} < e^{\frac{1}{2}\log(1+1/2n)} = \sqrt{\left(1+\frac{1}{2n}\right)} < 1+\frac{1}{4n}$$

Hence we have,

$$an^{n+\frac{1}{2}}e^{-n} < n! < an^{n+\frac{1}{2}}e^{-n}\left(1+\frac{1}{4n}\right).$$

There only remains for us to find the actual value of the limit a.

$$\sqrt{\pi} = \lim_{n\to\infty} \frac{(n!)^2\, 2^{2n}}{(2n)!\sqrt{n}}$$

Replacing $n!$ by $a_n n^{n+\cdots} e^{-n}$ and $(2n)!$ by $a_{2n} 2^{2n+ýÿ} n^{2n+ýÿ} e^{-2n}$, we immediately obtain,

$$\sqrt{\pi} = \lim_{n\to\infty} \frac{a_n^{\,2}}{a_{2n}\sqrt{2}} = \frac{a^2}{a\sqrt{2}},$$

whence $a = \sqrt{2\pi}$ The proof of Stirling's formula is thus complete.

In addition to its theoretical interest, Stirling's formula is a very useful tool for the numerical calculation of $n!$ when n is large. Instead of forming the product of a large number of integers, we have merely to calculate Stirling's expression using logarithms, which involves far fewer operations.

Thus, we obtain for $n = 10$ the value 3,598,696 from Stirling's expression, while the exact value is 3628800. The percentage error is barely 5/6 %.

9

Fundamental Theorem for Line Integrals

We had the Fundamental Theorem of Calculus that told us how to evaluate definite integrals. This told us,

$$\int_a^b F'(x)\,dx = F(b) - F(a)$$

It turns out that there is a version of this for line integrals over certain kinds of vector fields. Here it is.

Theorem: Suppose that C is a smooth curve given by $\vec{r}(t), a \le t \le b$.

Also suppose that f is a function whose gradient vector,, is continuous on C. Then,

$$\int_C \nabla f \bullet d\vec{r} = f(\vec{r}(b)) - f(\vec{r}(a))$$

Note that $\vec{r}(a)$ represents the initial point on C while $\vec{r}(b)$ represents the final point on C.

Also, we did not specify the number of variables for the function since it is really immaterial to the theorem. The theorem will hold regardless of the number of variables in the function.

Proof: This is a fairly straight forward proof.

For the purposes of the proof we'll assume that we're working in three dimensions, but it can be done in any dimension.

Let's start by just computing the line integral.

$$\int_C \nabla f \bullet d\vec{r} = \int_a^b \nabla f\left(\vec{r}\left(t\right)\right) \bullet \vec{r}'\left(t\right)$$

$$= \int_a^b \left(\frac{\partial f}{\partial x}\frac{dx}{dt} + \frac{\partial f}{\partial y}\frac{\partial y}{\partial t} + \frac{\partial f}{\partial z}\frac{dz}{dt} \right) dt$$

Now, at this point we can use the Chain Rule to simplify the integrand as follows,

$$\int_C \nabla f \bullet d\vec{r} = \int_a^b \left(\frac{\partial f}{\partial x}\frac{dx}{dt} + \frac{\partial f}{\partial y}\frac{\partial y}{\partial t} + \frac{\partial f}{\partial z}\frac{dz}{dt} \right) dt$$

$$= \int_a^b \frac{d}{dt}\left[f\left(\vec{r}\left(t\right)\right)\right] dt$$

To finish this off we just need to use the Fundamental Theorem of Calculus for single integrals.

$$\int_C \nabla f \bullet d\vec{r} = f\left(\vec{r}\left(b\right)\right) - f\left(\vec{r}\left(a\right)\right)$$

Let's take a quick look at an example of using this theorem.

Example: Evaluate $\int_C \nabla f \bullet d\vec{r}$ where

$$f\left(x, y, z\right) = \cos\left(\pi x\right) + \sin\left(\pi y\right) - xyz$$

and C is *any* path that starts $\left(1, \frac{1}{2}, 2\right)$ at and ends at $\left(2, 1, -1\right)$.

Solution: First let's notice that we didn't specify the path for getting from the first point to the second point. The reason for this is simple.

The theorem above tells us that all we need are the initial and final points on the curve in order to evaluate this kind of line integral.

So, let $\vec{r}\left(t\right), a \le t \le b$ be any path that starts at $\left(1, \frac{1}{2}, 2\right)$ and ends at $\left(2, 1, -1\right)$. Then,

$$\vec{r}(a)=\left\langle 1,\frac{1}{2},2\right\rangle$$

$$\vec{r}(b)=\langle 2,1,-1\rangle$$

Notice that we also didn't need the gradient vector to actually do this line integral. However, for the practice of finding gradient vectors here it is,

$$\nabla f=\langle -\pi sin(\pi x)-yz,\pi\cos(\pi y)-xz,-xy\rangle$$

The most important idea to get from this example is not how to do the integral as that's pretty simple, all we do is plug the final point and initial point into the function and subtract the two results. The important idea from this example (and hence about the Fundamental Theorem of Calculus) is that, for these kinds of line integrals, we didn't really need to know the path to get the answer. In other words, we could use any path we want and we'll always get the same results.

In the first section on line integrals (even though we weren't looking at vector fields) we saw that often when we change the path we will change the value of the line integral. We now have a type of line integral for which we know that changing the path will NOT change the value of the line integral.

Let's formalize this idea up a little. Here are some definitions. The first one we've already seen before, but it's been a while and it's important in this section so we'll give it again. The remaining definitions are new.

Definitions: First suppose that is a continuous vector field in some domain *D*.

1. $\vec{F}$ is a conservative vector field if there is a function *f* such that $\vec{F}=\nabla f$. The function *f* is called a potential function for the vector field. We first saw this definition in the first section of this chapter.

2. $\int_C \vec{F}\bullet d\vec{r}$ is independent of path if $\int_{C_1}\vec{F}\bullet d\vec{r}=\int_{C_2}\vec{F}\bullet d\vec{r}$

for any two paths C_1 and C_2 in D with the same initial and final points.

3. A path C is called closed if its initial and final points are the same point. For example a circle is a closed path.
4. A path C is simple if it doesn't cross itself. A circle is a simple curve while a figure 8 type curve is not simple.
5. A region D is open if it doesn't contain any of its boundary points.
6. A region D is connected if we can connect any two points in the region with a path that lies completely in D.
7. A region D is simply-connected if it is connected and it contains no holes. We won't need this one until the next section, but it fits in with all the other definitions given here so this was a natural place to put the definition.

With these definitions we can now give some nice facts.

Facts

1. $\int_C \nabla f \bullet d\vec{r}$ is independent of path.

   This is easy enough to prove since all we need to do is look at the theorem above. The theorem tells us that in order to evaluate this integral all we need are the initial and final points of the curve. This in turn tells us that the line integral must be independent of path.

2. $\vec{F}$ If is a conservative vector field then $\int_C \vec{F} \bullet d\vec{r}$ is independent of path.

   This fact is also easy enough to prove. If $\vec{F}$ is conservative then it has a potential function, f, and so the line integral becomes $\int_C \vec{F} \bullet d\vec{r} = \int_C \nabla f \bullet d\vec{r}$ .Then using the first fact we know that this line integral must be independent of path.

3. If $\vec{F}$ is a continuous vector field on an open connected region D and if $\int_C \vec{F} \bullet d\vec{r}$ is independent of path (for any path in D) then $\vec{F}$ is a conservative vector field on D.

4. If $\int_C \vec{F} \bullet d\vec{r}$ is independent of path then $\int_C \vec{F} \bullet d\vec{r} = 0$ for every closed path C.

5. If $\int_C \vec{F} \bullet d\vec{r} = 0$ for every closed path C then $\int_C \vec{F} \bullet d\vec{r}$ is independent of path.

These are some nice facts to remember as we work with line integrals over vector fields. Also notice that 2 & 3 and 4 & 5 are converses of each other.

CONSERVATIVE VECTOR FIELDS

In the previous section we saw that if we knew that the vector field $\vec{F}$ was conservative then $\int_C \vec{F} \bullet d\vec{r}$ was independent of path. This in turn means that we can easily evaluate this line integral provided we can find a potential function for $\vec{F}$.

In this section we want to look at two questions. First, given a vector field $\vec{F}$ is there any way of determining if it is a conservative vector field? Secondly, if we know that $\vec{F}$ is a conservative vector field how do we go about finding a potential function for the vector field?

The first question is easy to answer at this point if we have a two-dimensional vector field. For higher dimensional vector fields we'll need to wait until the final section in this chapter to answer this question. With that being said let's see how we do it for two-dimensional vector fields.

Theorem: Let $\vec{F} = P\vec{i} + Q\vec{j}$ be a vector field on an open and simply-connected region D. Then if P and Q have continuous first order partial derivatives in D and

$$\frac{\partial P}{\partial y} = \frac{\partial P}{\partial y}$$

the vector field $\vec{F}$ is conservative.

Let's take a look at a couple of examples.

Example: Determine if the following vector fields are conservative or not.

1. $\vec{F}(x,y) = \left(x^2 - yx\right)\vec{i} + \left(y^2 - xy\right)\vec{j}$
2. $\vec{F}(x,y) = \left(2xe^{xy} + x^2ye^{xy}\right)\vec{i} + \left(x^3e^{xy} + 2yxy\right)\vec{j}$

Solution: There really isn't too much to these. All we do is identify P and Q then take a couple of derivatives and compare the results.

1. $\vec{F}(x,y) = \left(x^2 - yx\right)\vec{i} + \left(y^2 - xy\right)\vec{j}$

In this case here is P and Q and the appropriate partial derivatives.

$$P = x^2 - yx \qquad \frac{\partial P}{\partial x} = -x$$

$$Q = y^2 - xy \qquad \frac{\partial Q}{\partial x} = -y$$

So, since the two partial derivatives are not the same this vector field is NOT conservative.

Example: Determine if the following vector fields are conservative and find a potential function for the vector field if it is conservative.

1. $\vec{F} = \left(2x^3y^4 + x\right)\vec{i} + \left(2x^4y^3 + y\right)\vec{j}$
2. $\vec{F}(x,y) = \left(2xe^{xy} + x^2ye^{xy}\right)\vec{i} + \left(x^3e^{xy} + 2y\right)\vec{j}$

Solution:

1. $$\vec{F} = \left(2x^3y^4 + x\right)\vec{i} + \left(2x^4y^3 + y\right)\vec{j}$$

Let's first identify P and Q and then check that the vector field is conservative..

$$P = 2x^3y^4 + x \qquad \frac{\partial P}{\partial y} = 8x^3y^3$$

$$Q = 2x^4y^3 + y \qquad \frac{\partial Q}{\partial x} = 8x^3y^3$$

So, the vector field is conservative. Now let's find the potential function. From the first fact above we know that,

$$\frac{\partial f}{\partial x} = 2x^3y^4 + x \qquad \frac{\partial f}{\partial y} = 2x^4y^3 + y$$

From these we can see that

$$f(x,y) = \int 2x^3y^4 + x\,dx \quad \text{or} \quad f(x,y) = \int 2x^4y^3 + y\,dy$$

We can use either of these to get the process started. Recall that we are going to have to be careful with the "constant of integration" which ever integral we choose to use. For this example let's work with the first integral and so that means that we are asking what function did we differentiate with respect to x to get the integrand.

This means that the "constant of integration" is going to have to be a function of y since any function consisting only of y and/or constants will differentiate to zero when taking the partial derivative with respect to x.

Here is the first integral.

$$f(x,y) = \int 2x^3y^4 + x\,dx$$

$$= \frac{1}{2}x^4y^4 + \frac{1}{2}x^2 + h(y)$$

where $h(y)$is the "constant of integration".

We now need to now determine $h(y)$. This is easier that it might at first appear to be. To get to this point we've used the fact that we knew P, but we will also need to use the fact that we know Q to complete the problem. Recall that Q is really the derivative of f with respect to y.

So, if we differentiate our function with respect to y we know what it should be.

So, let's differentiate f (including the $h(y)$) with respect to y and set it equal to Q since that is what the derivative is supposed to be.

$$\frac{\partial f}{\partial y} = 2x^4y^3 + h'(y) = 2x^4y^3 + y = Q$$

From this we can see that,

$$h'(y) = y$$

Notice that since $h'(y)$ is a function only of y so if there are any x's in the equation at this point we will know that we've made a mistake. At this point finding $h(y)$ is simple.

$$h(y)\int h'(y)dy = \int ydx = \frac{1}{2}y^2 + c$$

So, putting this all together we can see that a potential function for the vector field is,

$$f(x,y) = \frac{1}{2}x^4y^4 + \frac{1}{2}x^2 + \frac{1}{2}y^2 + c$$

Note that we can always check our work by verifying that $\nabla f \bullet \vec{F}$.

Also note that because the c can be anything there are an infinite number of possible potential functions, although they will only vary by an additive constant.

$$\vec{F}(x,y) = \left(2xe^{xy} + x^2ye^{xy}\right)\vec{i} + \left(x^3e^{xy} + 2y\right)\vec{j}$$

This one will go a lot faster since we don't need to go through as much explanation. We've already verified that this vector field is conservative in the first set of examples so we won't bother redoing that.

Let's start with the following,

$$\frac{\partial f}{\partial x} = 2xe^{xy} + x^2 ye^{xy}$$

$$\frac{\partial f}{\partial y} = x^2 e^{xy} + 2y$$

This means that we can do either of the following integrals,

$$f(x,y) = \int 2xe^{xy} + x^2 ye^{xy} dx \text{ or } f(x,y) = \int x^3 e^{xy} + 2y\,dy$$

While we can do either of these the first integral would be somewhat unpleasant as we would need to do integration by parts on each portion. On the other hand the second integral is fairly simple since the second term only involves y's and the first term can be done with the substitution $u = xy$. So, from the second integral we get,

$$f(x,y) = x^2 e^{xy} + y^2 + h(x)$$

Notice that this time the "constant of integration" will be a function of x. If we differentiate this with respect to x and set equal to P we get,

$$\frac{\partial f}{\partial x} = 2xe^{xy} + x^2 ye^{xy} + h'(x)$$

$$= 2xe^{xy} + x^2 ye^{xy} = P$$

So, in this case it looks like,

$$h'(x) = 0 \quad \Rightarrow \quad h(x) = c$$

So, in this case the "constant of integration" really was a constant. Sometimes this will happen and sometimes it won't.

Here is the potential function for this vector field.

$$f(x,y) = x^2 e^{xy} + y^2 + c$$

Now, as noted above we don't have a way (yet) of determining if a three-dimensional vector field is conservative or not. However, if we are given that a three-dimensional vector field is conservative finding a potential function is similar to the above process, although the work will be a little more involved.

In this case we will use the fact that,

$$\nabla f = \frac{\partial f}{\partial x}\vec{i} + \frac{\partial f}{\partial y}\vec{j} + \frac{\partial f}{\partial z}\vec{k} = P\vec{i} + Q\vec{j} + R\vec{k} = \vec{F}$$

Let's take a quick look at an example.

Example: Find a potential function for the vector field,

$$\vec{F} = 2xy^3z^4\vec{i} + 3x^2y^3z^3\vec{j} + 4x^2y^3z^3\vec{k}$$

Solution: We'll start off with the following equalities.

$$\frac{\partial f}{\partial x} = 2xy^3z^4$$

$$\frac{\partial f}{\partial y} = 3x^2y^2z^4$$

$$\frac{\partial f}{\partial z} = 4x^2y^3z^3$$

To get started we can integrate the first one with respect to x, the second one with respect to y, or the third one with respect to z. Let's integrate the first one with respect to x.

$$f(x,y,z) = \int 2xy^3z^4\,dx = x^2y^3z^4 + g(y,z)$$

Note that this time the "constant of integration" will be a function of both y and z since differentiating anything of that form with respect to x will differentiate to zero.

Now, we can differentiate this with respect to y and set it equal to Q. Doing this gives,

$$\frac{\partial f}{\partial y} = 3x^2y^2z^4 + g_y(y,z) = 3x^2y^2z^4 = Q$$

Of course we'll need to take the partial derivative of the constant of integration since it is a function of two variables. It looks like we've now got the following,

$$g_y(y,z) = 0$$

$$g(y,z) = h(z)$$

Since differentiating $g(y,z)$ with respect to y gives zero then $g(y,z)$ could at most be a function of z. This means that we now know the potential function must be in the following form.

$$f(x,y,z) = x^2y^3z^4 + h(z)$$

To finish this out all we need to do is differentiate with respect to z and set the result equal to R.

$$\frac{\partial f}{\partial z} = 4x^2y^3z^3 + h'(z) = 4x^2y^3z^3 = R$$

So,

$$h'(z) = 0 \quad \Rightarrow \quad h(z) = c$$

The potential function for this vector field is then,

$$f(x,y,z) = x^2y^3z^4 + c$$

Note that to keep the work to a minimum we used a fairly simple potential function for this example. It might have been possible to guess what the potential function was based simply on the vector field. However, we should be careful to remember that this usually won't be the case and often this process is required. Also, there were several other paths that we could have taken to find the potential function. Each would have gotten us the same result. Let's work one more slightly (and only slightly) more complicated example.

Example: Find a potential function for the vector field,

$$\vec{F} = \left(2x\cos(y) - 2z^3\right)\vec{i} + \left(3 + 2ye^z - x^2\sin(y)\right)\vec{j} + \left(y^2e^2 - 6xz^2\right)\vec{k}$$

Solution: Here are the equalities for this vector field.

$$\frac{\partial f}{\partial x} = 2x\cos(y) - 2z^3$$

$$\frac{\partial f}{\partial y} = 3 + 2ye^z \sin(y)$$

$$\frac{\partial f}{\partial z} = y^2e^2 - 6xz^2$$

For this example let's integrate the third one with respect to z.

$$f(x,y,z) = \int y^2e^z - 6xz^2 dz$$

$$= y^2e^z - 2xz^3 + g(x,y)$$

The "constant of integration" for this integration will be a function of both x and y.

Now, we can differentiate this with respect to x and set it equal to P. Doing this gives,

$$\frac{\partial f}{\partial x} = -2z^3 + g_x(x,y)$$

$$= 2x\cos(y) - 2z^3 = P$$

So, it looks like we've now got the following,

$$g_x(x,y) = 2x\cos(y) \quad \Rightarrow \quad g(x,y) = x^2\cos(y) + h(y)$$

The potential function for this problem is then,

$$f(x,y,z) = y^2e^z - 2xz^3 + x^2\cos(y) + h(y)$$

To finish this out all we need to do is differentiate with respect to y and set the result equal to Q.

$$\frac{\partial f}{\partial y} = 2ye^z - x^2\sin(y) + h'(y)$$

$$= 3 + 2ye^z - x^2\sin(y) = Q$$

So,

$$h'(y) = 3 \quad \Rightarrow \quad h(y) = 3y + c$$

The potential function for this vector field is then,

$$f(x,y,z) = y^2e^z - 2xz^3 + x^2\cos(y) + 3y + c$$

So, a little more complicated than the others and there are again many different paths that we could have taken to get the answer. We need to work one final example in this section.

Example: Evaluate $\int\limits_C \vec{F} \bullet d\vec{r}$ where

$$\vec{F} = \left(2x^3y^4 + x\right)\vec{i} + \left(2x^4y^3 + y\right)\vec{j}$$

and C is given by

$$\vec{r}(t) = \left(t\cos(\pi t) - 1\right)\vec{i} + \sin\left(\frac{\pi t}{2}\right)\vec{j}, 0 \le t \le 1$$

Solution: Now, we could use the techniques we discussed when we first looked at line integrals of vector fields however that would be particularly unpleasant solution.

Instead, let's take advantage of the fact that we know from Example above this vector field is conservative and that a potential function for the vector field is,

$$f(x,y)=\frac{1}{2}x^4y^4+\frac{1}{2}x^4+\frac{1}{2}y^2+c$$

Using this we know that integral must be independent of path and so all we need to do is use the theorem from the previous section to do the evaluation.

$$\int_C \vec{F}\bullet d\vec{r}=\int_C \nabla f\bullet d\vec{r}=f(\vec{r}(1))-f(\vec{r}(0))$$

where,

$$\vec{r}(1)=\langle -2,1\rangle$$
$$\vec{r}(0)=\langle -1,0\rangle$$

So, the integral is,

$$\begin{aligned}\int_C \vec{F}\bullet d\vec{r}&=f(-2,1)-f(-1,0)\\&=\left(\frac{21}{2}+c\right)-\left(\frac{1}{2}+c\right)\\&=10\end{aligned}$$

VECTOR FIELDS

We need to start this chapter off with the definition of a vector field as they will be a major component of both this chapter and the next. Let's start off with the formal definition of a vector field.

Definition: A vector field on two (or three) dimensionl space is a function $\vec{F}$ that assigns to each point (x,y) or (x,y,z) a two (or three dimensional) vector given by

$$\vec{F}(x,y)\text{ or }\vec{F}(x,y,z).$$

That may not make a lot of sense, but most people do know what a vector field is, or at least they've seen a sketch of a vector field. If you've seen a current sketch giving the direction and magnitude of a flow of a fluid or the direction and magnitude of the winds then you've seen a sketch of a vector field.

The standard notation for the function $\vec{F}$ is,

$$\vec{F}(x,y) = P(x,y)\vec{i} + Q(x,y)\vec{j}$$

$$\vec{F}(x,y,z) = P(x,y,z)i + Q(x,y,z)\vec{j} + R(x,y,z)\vec{k}$$

depending on whether or not we're in two or three dimensions. The function *P*, *Q*, *R* (if it is present) are sometimes called scalar functions.

Let's take a quick look at a couple of examples.

Example: Sketch each of the following direction fields.

1. $\vec{F}(x,y) = -y\vec{i} + x\vec{j}$

2. $\vec{F}(x,y,z) = 2x\vec{i} - 2y\vec{j} - 2x\vec{k}$

Solution:

1. $\vec{F}(x,y) = -y\vec{i} + x\vec{j}$

To graph the vector field we need to get some "values" of the function. This means plugging in some points into the function. Here are a couple of evaluations.

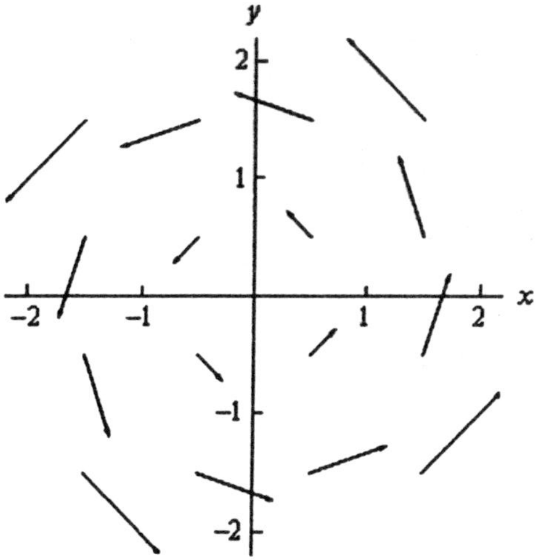

$$\vec{F}\left(\frac{1}{2},\frac{1}{2}\right) = -\frac{1}{2}\vec{i} + \frac{1}{2}\vec{j}$$

$$F\left(\frac{1}{2},-\frac{1}{2}\right)=-\left(-\frac{1}{2}\right)\vec{i}+\frac{1}{2}\vec{j}=\frac{1}{2}\vec{i}+\frac{1}{2}\vec{j}$$

$$\vec{F}\left(\frac{3}{2},\frac{1}{4}\right)=-\frac{1}{4}\vec{i}+\frac{3}{2}\vec{j}$$

So, just what do these evaluations tell us? Well the first one tells us that at the point $\left(\frac{1}{2},\frac{1}{2}\right)$ we will plot the vector $-\frac{1}{2}\vec{i}+\frac{1}{2}\vec{j}$. Likewise, the third evaluation tells us that at the point $\left(\frac{3}{2},\frac{1}{4}\right)$ we will plot the vector $-\frac{1}{4}\vec{i}+\frac{3}{2}\vec{j}$.

We can continue in this fashion plotting vectors for several points and we'll get the following sketch of the vector field.

If we want significantly more points plotted then it is usually best to use a computer aided graphing system such as Maple or Mathematica. Here is a sketch with many more vectors included that was generated with Mathematica.

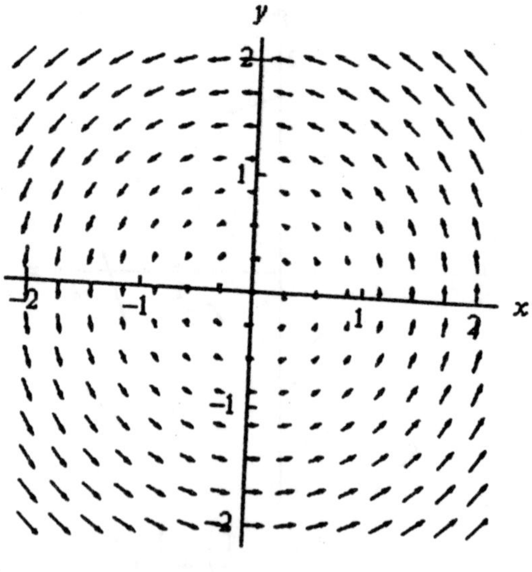

2. $\vec{F}(x,y,z)=2x\vec{i}-2y\vec{j}-2x\vec{k}$

In the case of three dimensional vector fields it is almost always better to use Maple, Mathematica, or some other such tool. Despite that let's go ahead and do a couple of evaluations anyway.

$$\vec{F}(1,-3,2) = 2\vec{i} + 6\vec{j} - 2\vec{k}$$

$$\vec{F}(0,5,3) = -10j$$

Notice that z only affect the placement of the vector in this case and does not affect the direction or the magnitude of the vector. Sometimes this will happen so don't get excited about it when it does. Recall that given a function $f(x,y,z)$ the gradient vector is defined by,

$$\nabla f = \langle f_x, f_y, f_z \rangle$$

This is a vector field and is often called a gradient vector field. In these cases the function $f(x,y,z)$ is often called a scalar function to differentiate it from the vector field.

Example: Find the gradient vector field of the following functions.

1. $f(x,y) = x^2 \sin(5y)$
2. $f(x,y,z) = ze^{-xy}$

Solution:

1. $f(x,y) = x^2 \sin(5y)$

Note that we only gave the gradient vector definition for a three dimensional function, but don't forget that there is also a two dimension definition. All that we need to drop off the third component of the vector.

Here is the gradient vector field for this function.

$$\nabla f = \langle 2x\sin(5y), 5x^2 \cos(5y) \rangle$$

2. $f(x,y,z) = ze^{-xy}$

There isn't much to do here other than take the gradient.

$$\nabla f = \langle -yze^{-xy}, -xze^{-xy}, e^{-xy} \rangle$$

Let's do another example that will illustrate the relationship between the gradient vector field of a function and its contours.

Example: Sketch the gradient vector field for $f(x, y) = x^2 + y^2$ as well as several contours for this function.

Solution: Recall that the contours for a function are nothing more than curves defined by,

$$f(x, y) = k$$

for various values of k. So, for our function the contours are defined by the equation,

$$x^2 + y^2 = k$$

and so they are circles centered at the origin with radius $\sqrt{k}$.

Here is the gradient vector field for this function.

$$\nabla f(x, y) = 2xi + 2yj$$

Here is a sketch of several of the contours as well as the gradient vector field.

Notice that the vectors of the vector field are all perpendicular (or orthogonal) to the contours. This will always be the case when we are dealing with the contours of a function as well as its gradient vector field. The k's we used for the graph above were 1.5, 3, 4.5, 6, 7.5, 9, 10.5, 12, and 13.5.

Now notice that as we increased k by 1.5 the contour curves get closer together and that as the contour curves get closer together the larger vectors become.

In other words, the closer the contour curves are (as k is increased by a fixed amount) the faster the function is changing at that point. Also recall that the direction of fastest change for a function is given by the gradient vector at that point. Therefore, it should make sense that the two ideas should match up as they do here.

The final topic of this section is that of conservative vector fields. A vector field $\vec{F}$ is called a conservative vector field if there exists a function f such that $\vec{F} = \nabla f$. If $\vec{F}$ is a conservative vector field then the function, f, is called a potential function for $\vec{F}$.

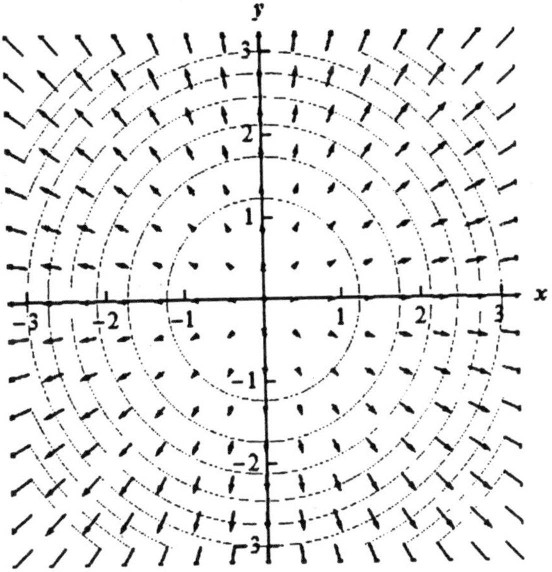

All this definition is saying is that a vector field is conservative if it is also a gradient vector field for some function. For instance the vector field $\vec{F} = y\vec{i} + x\vec{j}$ is a conservative vector field with a potential function of $f(x,y) = xy$ because $\nabla f = \langle y, x \rangle$.

On the other hand, $\vec{F} = -y\vec{i} + x\vec{j}$ is not a conservative vector field since there is no function *f* such that $\vec{F} = \nabla f$ .If you're not sure that you believe this at this point be patient, we will be able to prove this in a couple of sections. In that section we will also show how to find the potential function for a conservative vector field.

LINE INTEGRALS PART I

In this section we are now going to introduce a new kind of integral. However, before we do that it is important to note that you will need to remember how to parameterize equations, or put another way, you will need to be able to write down a set of parametric equations for a given curve. If you need some review you should go back and review some of the basics of parametric equations and curves. Here are some of the more basic curves that we'll need to know how to do as well as limits on the parameter if they are required.

Curve	**Parametric**	**Equations**
(Ellipse) $\dfrac{x^2}{a^2}+\dfrac{y^2}{b^2}=1$	Counter-Clockwise $x = a\cos(t)$ $y = b\sin(t)$ $0 \le t \le 2\pi$	Clockwise $x = a\cos(t)$ $y = -b\sin(t)$ $0 \le t \le 2\pi$
(Circle) $x^2 + y^2 = r^2$	Counter-Clockwise $x = r\cos(t)$ $y = r\sin(t)$ $0 \le t \le 2\pi$	Clockwise $x = r\cos(t)$ $y = -r\sin(t)$ $0 \le t \le 2\pi$
$y = f(x)$	$x = t$ $y = f(t)$	
$x = g(y)$	$y = g(t)$ $y = t$	
Line segment from (x_0, y_0, z_0) to (x_1, y_1, z_1)	$x = (1-t)x_0 + tx_1$ $y = (1-t)y_0 + ty_1, 0 \le t \le 1$ $z = (1-t)z_0 + tz_1$	

With the final one we gave both the vector form of the equation as well as the parametric form and if we need the two-dimensional version then we just drop the *z* components.

In fact, we will be using the two-dimensional version of this in this section.

For the ellipse and the circle we've given two parameterizations, one tracing out the curve clockwise and the other counter-clockwise. As we'll eventually see the direction that the curve is traced out can, on occasion, change the answer. Also, both of these "start" on the positive *x*-axis at $t = 0$.

Now let's move on to line integrals $f(x)$. We integrated, a function of a single variable, over an interval $[a,b]$. In this case we were thinking of x as taking all the values in this interval starting at a and ending at b. With line integrals we will start with integrating the function $f(x,y)$, a function of two variables, and the values of x and y that we're going to use will be the points, (x,y), that lie on a curve C.

Note that this is different from the double integrals that we were working with in the previous chapter where the points came out of some two-dimensional region. Let's start with the curve C that the points come from. We will assume that the curve is *smooth* (defined shortly) and is given by the parametric equations,

$$x = h(t) \quad y = g(t) \quad a \le t \le b$$

We will often want to write the parameterization of the curve as a vector function. In this case the curve is given by,

$$\vec{r}(t) = h(t)\vec{i} + g(t)\vec{j} \quad a \le t \le b$$

The curve is called smooth if $\vec{r}(t)$ is continuous and $\vec{r}(t) \neq 0$ for all t. The line integral of $f(x,y)$ along C is denoted by,

$$\int_C f(x,y)\,ds$$

We use a *ds* here to acknowledge the fact that we are moving along the curve, C, instead of the x-axis (denoted by dx) or the y-axis (denoted by dy). Because of the *ds* this is sometimes called the line integral of f with respect to arc length. We've seen the notation *ds* before. When we looked at the arc length of a curve given by parametric equations we found it to be,

$$L = \int_a^b ds, \quad where\ ds = \sqrt{\left(\frac{dx}{dt}\right)^2 + \left(\frac{dy}{dt}\right)^2}\,dt$$

It is no coincidence that we use *ds* for both of these problems. The *ds* is the same for both the arc length integral and the notation for the line integral. So, to compute a line integral we will convert everything over to the parametric equations. The line integral is then,

$$\int_C f(x,y)\,ds = \int_a^b f(h(t),g(t))\sqrt{\left(\frac{dx}{dt}\right)^2+\left(\frac{dy}{dt}\right)^2}\,dt$$

Don't forget to plug the parametric equations into the function as well. If we use the vector form of the parameterization we can simplify the notation up somewhat by noticing that,

$$\sqrt{\left(\frac{dx}{dt}\right)^2+\left(\frac{dy}{dt}\right)^2} = \|\vec{r}'(t)\|$$

where $\|\vec{r}(t)\|$ is the magnitude or norm of $\vec{r}(t)$. Using this notation the line integral becomes,

$$\int_C f(x,y)\,ds = \int_a^b f(h(t),g(t))\|\vec{r}(t)\|dt$$

Note that as long as the parameterization of the curve C is traced out exactly once at t increases from a to b the value of the line integral will be independent of the parameterization of the curve.

Example: Evaluate $\int_C xy^4\,ds$ where C is the right half of the circle, $x^2+y^2=16$. rotated in the counter clockwise direction.

Solution: We first need a parameterization of the circle. This is given by,

$$x = 4\cos t$$

$$y = 4\sin t$$

We now need a range of t's that will give the right half of the circle. The following range of t's will do this.

$$-\frac{\pi}{2} \le t \le \frac{\pi}{2}$$

Now, we need the derivatives of the parametric equations and let's compute ds.

$$\frac{dx}{dt} = -4\sin t$$

$$\frac{dy}{dt} = 4\cos t$$

$$ds = \sqrt{16\sin^2 t + 16\cos^2 t}\, dt = 4dt$$

The line integral is then,

$$\int_C xy^4\, ds = \int_{-\pi/2}^{\pi/2} 4\cos t(4\sin t)^4 (4)dt$$

$$= 4096 \int_{-\pi/2}^{\pi/2} \cos t \sin^4 t dt$$

$$= \frac{4096}{5} \sin^5 t \Big|_{-\frac{\pi}{2}}^{\frac{\pi}{2}}$$

$$= \frac{8192}{5}$$

Next we need to talk about line integrals over piecewise smooth curves.A piecewise smooth curve is any curve that can be written as the union of a finite number of smooth curves, $C_1,...,C_n$ where the end point of is the starting point of C_t. Below is an illustration of a piecewise smooth curve.

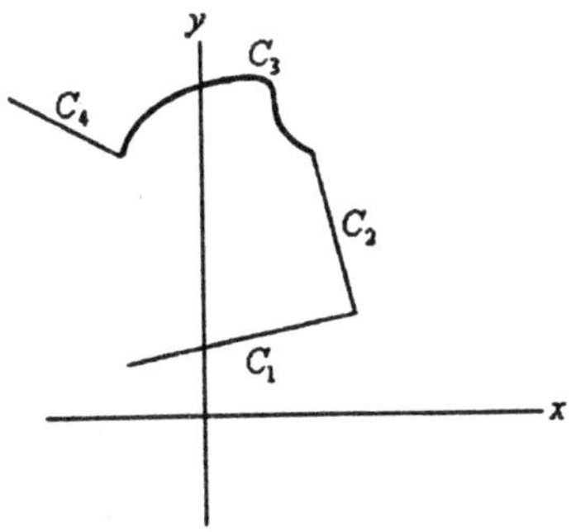

Evaluation of line integrals over piecewise smooth curves is a relatively simple thing to do. All we do is evaluate the line integral over each of the pieces and then add them up. The line integral for some function over the above piecewise curve would be,

$$\int_C f(x,y)\,ds = \int_{C_1} f(x,y)\,ds + \int_{C_2} f(x,y)\,ds + \int_{C_3} f(x,y)\,ds + \int_{C_4} f(x,y)\,ds$$

Let's see an example of this.

Example: Evaluate where $\int_C 4x^3 ds$ C is the curve shown below.

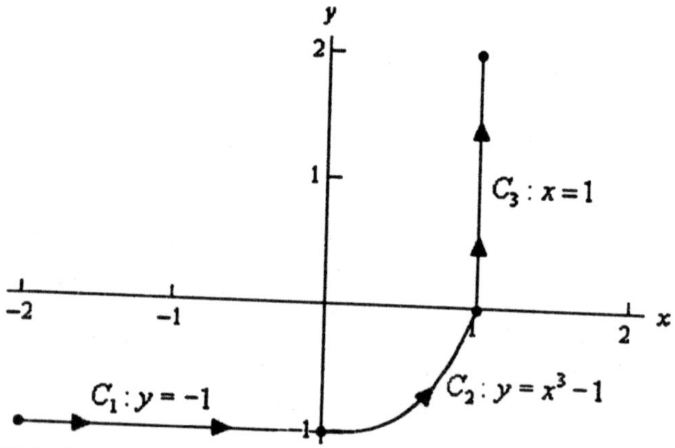

Solution: So, first we need to parameterize each of the curves.

$$C_1 : \ x = t,\ y = -1, \quad -2 \le t \le 0$$

$$C_2 : \ x = t,\ y = t^3 - 1, \quad 0 \le t \le 1$$

$$C_3 : \ x = t,\ y = t, \quad 0 \le t \le 2$$

Now let's do the line integral over each of these curves.

$$\int_{C_1} 4x^3 ds = \int_{-2}^{0} 4t^2 \sqrt{(1)^2 + (0)^2}\, ds$$

$$= \int_{-2}^{0} 4t^3 dt = t^4 \Big|_{-2}^{0} = -16$$

$$\int_{C_2} 4x^3\, ds = \int_0^1 4t^3 \sqrt{(1)^2 + (3t^2)^2}\, dt$$

$$= \int_0^1 4t^3 \sqrt{1 + 9t^4}\, dt$$

$$= \frac{1}{9}\left(\frac{2}{3}\right)(1 + 9t^4)\Big|_0^1 = \frac{2}{27}\left(10^{\frac{3}{2}} - 1\right) = 2.268$$

Finally, the line integral that we were asked to compute is,

$$\int_C 4x^3\, ds = \int_{C_1} 4x^3\, ds + \int_{C_2} 4x^3\, ds + \int_{C_3} 4x^3\, ds$$

$$= -16 + 2.268 + 8$$

$$= -5.732$$

Notice that we put direction arrows on the curve in the above example. The direction of motion along a curve *may* change the value of the line integral as we will see in the next section. Also note that the curve can be thought of a curve that takes us from the point $(-2,-1)$ to the point $(2,1)$. Let's first see what happens to the line integral if we change the path between these two points.

Example: Evaluate were $\int_C 4x^3\, ds$ is the line segment from $(-2,-1)$ to $(2,1)$.

Solution: From the parameterization formulas at the start of this section we know that the line segment start at $(-2,-1)$ and ending at $(2,1)$ is given by,

$$\vec{r}(t) = (1-t)\langle -2,-1\rangle + t\langle 1,2\rangle$$

$$= \langle -2+3t, -1+3t\rangle$$

for $0 \le t \le 1$. This means that the individual parametric equations are,

$$x = -2 + 3t$$

$$y = -1 + 3t$$

Using this path the line integral is,

$$\int_C 4x^3\,ds = \int_0^1 4(-2+3t)^3\sqrt{9+9}\,dt$$

$$= 12\sqrt{2}\left(\frac{1}{12}\right)(-2+3t)^4\Big|_0^1$$

$$= 12\sqrt{2}\left(-\frac{5}{4}\right)$$

$$= -15\sqrt{2} = -21.213$$

When doing these integrals don't forget simple substitutions to avoid having to do things like cubing out that term. Cubing it out is not that difficult, but is more work than a simple substitution.

So, the previous two examples seem to suggest that if we change the path between two points then the value of the line integral (with respect to arc length) will change.

While this will happen fairly regularly we can't assume that it will always happen. In a later section we will investigate this idea in more detail

Example: Evaluate $\int_C 4x^3 ds$ were C is the line segment from $(-2,-1)$ to $(2,1)$.

Solution: This one isn't much different, work wise, from the previous example. Here is the parameterization of the curve.

$$\vec{r}(t) = (1-t)\langle 1,2\rangle + t\langle -2,-1\rangle$$
$$= \langle 1-3t, 2-3t\rangle$$

for $0 \le t \le 1$. Remember that we are switch the direction of the curve and this will also change the parameterization so we can make sure that we start/end at the proper point.

Here is the line integral.

$$\int_C 4x^3 ds = \int_0^1 4(1-3t)^3\sqrt{9+9}$$

$$=12\sqrt{2}\left(\frac{1}{12}\right)(-2+3t)^4\Big|_0^1$$

$$=12\sqrt{2}\left(-\frac{5}{4}\right)$$

$$=-15\sqrt{2}=-21.213$$

$$=12\sqrt{2}\left(\frac{1}{12}\right)(-2+3t)^4\Big|_0^1$$

$$=12\sqrt{2}\left(-\frac{5}{4}\right)$$

$$=-15\sqrt{2}=-21.213$$

So, it looks like when we switch the direction of the curve the line integral (with respect to arc length) will not change. This will always be true for these kinds of line integrals. However, there are other kinds of line integrals in which this won't be the case. We will see more examples of this in the next couple of sections so don't get it into your head that changing the direction will never change the value of the line integral.

Before working another example let's formalize this idea up somewhat. Let's suppose that the curve C has the parameterization $x = h(t), y = g(t)$. Let's also suppose that the initial point on the curve is A and the final point on the curve is B. The parameterization $x = h(t), y = g(t)$ will then determine an orientation for the curve where the positive direction is the direction that is traced out as t increases. Finally, let $-C$be the curve with the same points as C, however in this case the curve has B as the initial point and A as the final point, again t is increasing as we traverse this curve. In other words, given a curve C, the curve $-C$is the same curve as C except the direction has been reversed.

We then have the following fact about line integrals with respect to arc length.

Fact

$$\int_{C} f(x,y)\,ds = \int_{-C} f(x,y)\,ds$$

So, for a line integral with respect to arc length we can change the direction of the curve and not change the value of the integral. This is a useful fact to remember as some line integrals will be easier in one direction than the other.

Example: Evaluate for each of the following curves.

1. $C_1 : y = x^2, -1 \le x \le 1$
2. C_2 : The line segment from $(-1,1)$ to $(1,1)$.
3. C_3: The line segment from $(1,1)$ to $(-1,1)$.

Solution: Before working any of these line integrals let's notice that all of these curves are paths that connect the points $(-1,1)$ and $(1,1)$. Also notice that $C_3 = -C_2$ and so by the fact above these two should give the same answer. Here is a sketch of the three curves and note that the curves illustrating and have been separated a little to show that they are separate curves in some way even thought they are the same line.

1. $C_1 : y = x^2, -1 \le x \le 1$

Here is a parameterization for this curve.

$$C_1 : x = t, y = t^2, -1 \le t \le 1$$

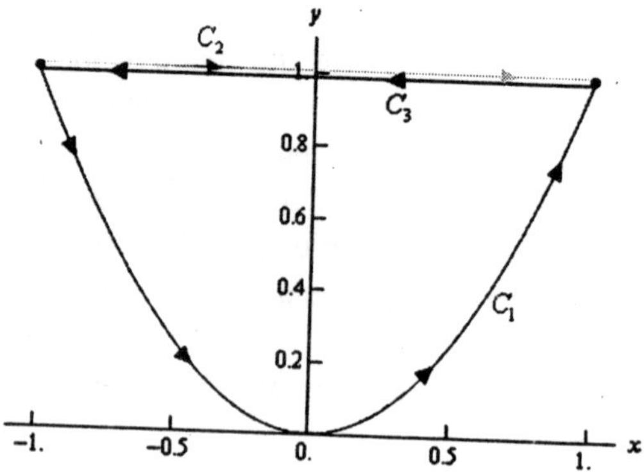

Here is the line integral.

$$\int_C x\,ds = \int_{-1}^{1} t\sqrt{1+4t^2}\,dt = \frac{1}{12}\left(1+4t^2\right)^{\frac{3}{2}}\Bigg|_{-1}^{1} = 0$$

2. C_2 : The line segment from $(-1,1)$ to $(1,1)$.

There are two parameterizations that we could use here for this curve. The first is to use the formula we used in the previous couple of examples. That parameterization is,

$$\begin{aligned} C_2 : \vec{r}(t) &= (1-t)\langle -1,1\rangle + t\langle 1,1\rangle \\ &= \langle 2t-1, 1\rangle \end{aligned}$$

for $0 \le t \le 1$.

Sometimes we have no choice but to use this parameterization. However, in this case there is a second (probably) easier parameterization. The second one uses the fact that we are really just graphing a portion of the line $y=1$. Using this the parameterization is,

$$C_2 : x = t, y = 1, -1 \le t \le 1$$

This will be a much easier parameterization to use so we will use this. Here is the line integral for this curve.

$$\int_C x\,ds = \int_{-1}^{1} t\sqrt{1+0}\,dt = \frac{1}{2}t^2\Bigg|_{-1}^{1} = 0$$

Note that this time, unlike the line integral we worked with in Examples we got the same value for the integral despite the fact that the path is different. This will happen on occasion. We should also not expect this integral to be the same for all paths between these two points. At this point all we know is that for these two paths the line integral will have the same value. It is completely possible that there is another path between these two points that will give a different value for the line integral.

3. C_3: The line segment from $(1,1)$ to $(-1,1)$.

Now, according to our fact above we really don't need to do

anything here since we know that $C_3 = -C_2$. The fact tells us that this line integral should be the same as the second part (*i.e.* zero). However, let's verify that, plus there is a point we need to make here about the parameterization.

Here is the parameterization for this curve.

$$C_3 : \vec{r}(t) = (1-t)\langle 1,1\rangle + t\langle -1,1\rangle$$
$$= \langle 1-2t, 1\rangle$$

for $0 \le t \le 1$.

Note that this time we can't use the second parameterization that we used in part (b) since we need to move from right to left as the parameter increases and the second parameterization used in the previous part will move in the opposite direction.

Here is the line integral for this curve.

$$\int_{C_3} x\,ds = \int_{-1}^{1} (1-2t)\sqrt{4+0}\,dt = 2\left(t-t^2\right)\Big|_0^1 = 0$$

Sure enough we got the same answer as the second part.

To this point in this section we've only looked at line integrals over a two-dimensional curve. However, there is no reason to restrict ourselves like that. We can do line integrals over three-dimensional curves as well. Let's suppose that the three-dimensional curve C is given by the parameterization,

$$x = x(t)$$
$$y = y(t)$$
$$z = z(t)$$
$$z \le t \le b$$

then the line integral is given by,

$$\int_C f(x,y,z)\,ds = \int_a^b f(x(t),y(t),z(t))$$

$$\sqrt{\left(\frac{dx}{dt}\right)^2 + \left(\frac{dy}{dt}\right)^2 + \left(\frac{dz}{dt}\right)^2}\,dt$$

Note that often when dealing with three-dimensional space the parameterization will be given as a vector function.

$$\vec{r}(t) = \langle x(t), y(t), z(t) \rangle$$

Notice that we changed up the notation for the parameterization a little. Since we rarely use the function names we simply kept the *x*, *y*, and *z* and added on the (t)part to denote that they may be functions of the parameter.

Also notice that, as with two-dimensional curves, we have,

$$\sqrt{\left(\frac{dx}{dt}\right)^2 + \left(\frac{dy}{dt}\right)^2 + \left(\frac{dz}{dt}\right)^2} = \left\|\vec{r}'(t)\right\|$$

and the line integral can again be written as,

$$\int_C f(x, y, z)\, ds = \int_a^b f(x(t), y(t), z(t)) \left\|\vec{r}'(t)\right\| dt$$

So, outside of the addition of a third parametric equation line integrals in three-dimensional space work the same as those in two-dimensional space. Let's work a quick example.

Example: Evaluate $\int_C xyz\, ds$ where C is the helix given by $\left\|\vec{r}(t)\right\| = \langle \cos(t), \sin(t), 3t \rangle$, $0 \le t \le 4\pi$.

Solution: Note that we first saw the vector equation for a helix back in the Vector Functions section. Here is a quick sketch of the helix.

Here is the line integral.

$$\int_C xyz\, ds = \int_0^{4\pi} 3t\cos(t)\sin(t)\sqrt{\sin^2 t + \cos^2 t + 9}dt$$

$$= \int_0^{4\pi} 3t\left(\frac{1}{2}\sin(2t)\right)\sqrt{1+9}$$

$$= \frac{3\sqrt{10}}{2}\left(\frac{1}{4}\sin(2t) - \frac{t}{2}\cos(2t)\right)\Bigg|_0^{4\pi}$$

$$= -3\sqrt{10\pi}$$

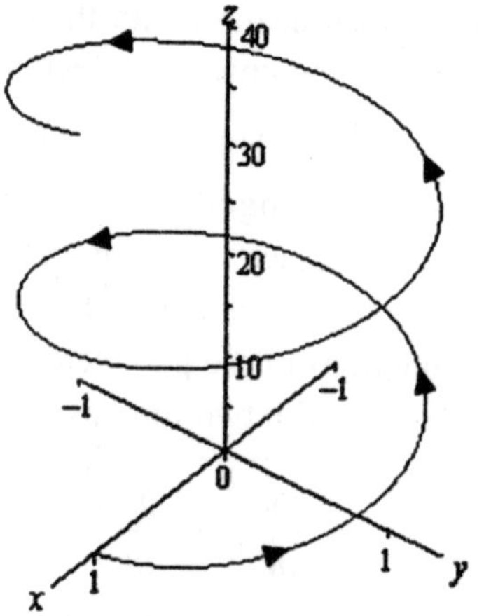

You were able to do that integral right? It required integration by parts.

LINE INTEGRALS PART II

In the previous section we looked at line integrals with respect to arc length. In this section we want to look at line integrals with respect to x and/or y. As with the last section we will start with a two-dimensional curve C with parameterization,

$$x = x(t)$$

$$y = y(t)$$

$$a \leq t \leq b$$

The line integral of f with respect to x is,

$$\int_C f(x,y)\,dx = \int_a^b f(x(t),y(t))\,x'(t)\,dt$$

The line integral of f with respect to y is,

$$\int_C f(x,y)\,dy = \int_a^b f(x(t),y(t))\,y'(t)\,dt$$

Note that the only notational difference between these two and the line integral with respect to arc length (from the previous section) is the differential. These have a dx or dy while the line integral with respect to arc length has a ds. So when evaluating line integrals be careful to first note which differential you've got so you don't work the wrong kind of line integral. These two integral often appear together and so we have the following shorthand notation for these cases.

$$\int_C P\,dx + Q\,dy = \int_a^b P(x,y)\,dx + \int_C Q(x,y)\,dy$$

Let's take a quick look at an example of this kind of line integral.

Example: Evaluate $\int_C \sin(\pi y)\,dy + yx^2\,dx$ where C is the line segment from $(0,2)$ to $(1,4)$.

Solution: Here is the parameterization of the curve.

$$\vec{r}(t) = (1-t)\langle 0,2\rangle + t\langle 1,4\rangle = \langle t, 2+2t\rangle \qquad 0 \le t \le 1$$

The line integral is,

$$\begin{aligned}
\int_C \sin(\pi y)\,dy + yx^2\,dx &= \int_C \sin(\pi y)\,dy + \int_C yx^2\,dx \\
&= \int_0^1 \sin(\pi(2+2t))(2)\,dt - \int_0^1 (2+2t)(t)^2(1) \\
&= -\frac{1}{\pi}\cos(2\pi - 2\pi t)\Big|_0^1 + \left(\frac{2}{3}t^3 + \frac{1}{2}t^4\right)\Big|_0^1 \\
&= \frac{7}{6}
\end{aligned}$$

In the previous section we saw that changing the direction of the curve for a line integral with respect to arc length doesn't change the value of the integral. Let's see what happens with line integrals with respect to x and/or y.

Example: Evaluate $\int_C \sin(\pi y)\,dy + yx^2\,dx$ where C is the line segment from $(1,4)$ to $(0,2)$.

Solution: So, we simply changed the direction of the curve. Here is the new parameterization.

$$\vec{r}(t) = (1-t)\langle 1,4\rangle + t\langle 0,2\rangle = \langle 1-t, 4-2t\rangle \quad 0 \le t \le 1$$

The line integral in this case is,

$$\int_C \sin(\pi y)\,dy + yx^2\,dx$$

$$= \int_C \sin(\pi y)\,dy + \int_C yx^2\,dx - \int_0^1 \sin(\pi(4-2t))(-2)\,dt$$

$$+ \int_0^1 (4-2t)(1-t)^2(-1)\,dt$$

$$= -\frac{1}{\pi}\cos(4\pi - 2\pi t)\Big|_0^1 - \left(-\frac{1}{2}t^4 + \frac{8}{3}t^3 - 5t^2 + 4t\right)\Big|_0^1$$

$$= -\frac{7}{6}$$

So, switching the direction of the curve got us a different value or at least the opposite sign of the value from the first example. In fact this will always happen with these kinds of line integrals.

Fact

If C is any curve then,

$$\int_{-C} f(x,y)\,dx = -\int_C f(x,y)\,dx \quad \text{and}$$

$$\int_{-C} f(x,y)\,dx = -\int_C f(x,y)\,dy$$

With the combined form of these two integrals we get,

$$\int_{-C} P\,dx + Q\,dy = -\int_C P\,dx + Q\,dy$$

We can also do these integrals over three-dimensional curves as well. In this case we will pick up a third integral (with respect to z) and the three integrals will be.

$$\int_C f(x,y,z)\,dx = \int_a^b f(x(t), y(t), z(t))\,x'(t)\,dt$$

$$\int_C f(x,y,z)\,dy = \int_a^b f(x(t),y(t),z(t))\,y'(t)\,dt$$

$$\int_C f(x,y,z)\,dz = \int_a^b f(x(t),y(t),z(t))\,z'(t)\,dt$$

where the curve C is parameterized by

$$x = x(t)$$
$$y = y(t)$$
$$z = z(t)$$
$$a \le t \le b$$

As with the two-dimensional version these three will often occur together so the shorthand we'll be using here is,

$$\int_{-C} P\,dx + Q\,dy + R\,dz = \int_{-C} P(x,y,z)\,dx + \int_{-C} Q(x,y,z)\,dy + \int_{-C} R(x,y,z)\,dz$$

Let's work an example.

Example: Evaluate $\int_{-C} y\,dx + x\,dy + z\,dz$ where C is given

by $x = \cos t, y = \sin t,\ z = t^2, 0 \le t \le 2\pi$

Solution: So, we already have the curve parameterized so there really isn't much to do other than evaluate the integral.

$$\begin{aligned}\int_C f\,dx + x\,dy + z\,dz &= \int_C f\,dx + \int_C x\,dy + \int_C z\,dz \\ &= \int_0^{2\pi} \sin t(-\sin t)\,dt + \int_0^{2\pi} \cos t(\cos t)\,dt \\ &\quad + \int_0^{2\pi} t^2(2t)\,dt - \int_0^{2\pi} \sin^2 t\,dt \\ &\quad + \int_0^{2\pi} \cos^2 t\,dt + \int_0^{2\pi} 2t^3\,dt\end{aligned}$$

$$= -\frac{1}{2}\int_0^{2\pi}\left(1-\cos(2t)\right)dt + \frac{1}{2}\int_0^{2\pi}\left(1+\cos(2t)\right)dt$$

$$+\int_0^{2\pi} 2t^3 dt$$

$$= \left(-\frac{1}{2}\left(t-\frac{1}{2}\sin(2t)\right)\right.$$

$$\left.+\frac{1}{2}\left(t+\frac{1}{2}\sin(2t)\right)+\frac{1}{2}t^4\right)\Bigg|_0^{2\pi}$$

$$= 8\pi^4$$

LINE INTEGRALS OF VECTOR FIELDS

In the previous two sections we looked at line integrals of functions. In this section we are going to evaluate line integrals of vector fields. We'll start with the vector field,

$$\vec{F}(x,y,z) = P(x,y,z)\vec{i} + Q(x,y,z)\,j + R(x,y,z)\vec{k}$$

and the three-dimensional, smooth curve given by

$$\vec{r}(t) = x(t)\vec{i} + y(\vec{j}) + z(t)\vec{k} \quad a \le t \le b$$

The line integral of $\vec{F}$ along C is

$$\int_C \vec{F}\bullet d\vec{r} = \int_a^b \vec{F}(\vec{r}(t))\bullet\vec{r}(t)\,dt$$

Note the notation in the left side. That really is a dot product of the vector field and the differential and the differential really is a vector. Also, $\vec{F}(\vec{r}(t))$ is a shorthand for,

$$\vec{F}(\vec{r}(t)) = \vec{F}(x(t), y(t), z(t))$$

We can also write line integrals of vector fields as a line integral with respect to arc length as follows,

$$\int_C \vec{F}\bullet d\vec{r} = \int_a^b \vec{F}\bullet\vec{T}\,ds$$

here $\vec{T}(t)$ is the unit tangent vector and is given by,

$$\vec{T}(t)=\frac{\vec{r}'(t)}{\left\|\vec{r}'(t)\right\|}$$

If we use our knowledge on how to compute line integrals with respect to arc length we can see that this second form is equivalent to the first form given above.

$$\int_C \vec{F}\bullet d\vec{r}=\int_C \vec{F}\bullet\vec{T}\,ds$$

$$=\int_a^b \vec{F}(\vec{r}(t))\bullet\frac{\vec{r}'(t)}{\left\|\vec{r}'(t)\right\|}\left\|\vec{r}'(t)\right\|dt$$

$$=\int_a^b \vec{F}(\vec{r}(t))\bullet\vec{r}'(t)\,dt$$

In general we use the first form to compute these line integrals as it is usually much easier to use. Let's take a look at a couple of examples.

Example: Evaluate $\int_C \vec{F}\bullet d\vec{r}$ where

$$\vec{F}(x,y,z)=8x^2yz\vec{i}+5z\vec{i}-4xy\vec{k}$$

and C is the curve given by

$$\vec{r}(t)=t\vec{i}+t^2\vec{j}+t^3\vec{k},0\le t\le 1.$$

Solution: We first need the vector field evaluated along the curve.

$$\vec{F}(\vec{r}(t))=8t^2\left(t^2\right)\left(t^3\right)\vec{i}+5t^3\vec{j}-4t\left(t^2\right)\vec{k}=8t^7\vec{i}+5t^3\vec{j}-4t^3\vec{k}$$

Next we need the derivative of the parameterization.

$$\vec{r}(t)=\vec{i}+2t\,\vec{j}+3t^2\vec{k}$$

Finally, let's get the dot product taken care of.

$$\vec{F}(\vec{r}(t))\bullet\vec{r}'(t)=8t^7+10t^4-12t^5$$

The line integral is then,

$$\int_C \vec{F}\bullet d\vec{r}=\int_0^1 8t^7+10t^4-12t^5\,dt$$

$$= \left(t^8 + 2t^5 - 2t^6\right)\Big|_0^1$$

$$= 1$$

Example: Evaluate $\int\limits_C \vec{F} \bullet d\vec{r}$ where

$$\vec{F}(x, y, z) = xz\vec{i} - yz\vec{k}$$

and C is the line segment from $(-1,2,0)$ and $(3,0,1)$.

Solution: We'll first need the parameterization of the line segment. We saw how to get the parameterization of line segments in the first section on line integrals. We've been using the two dimensional version of this over the last couple of sections. Here is the parameterization for the line.

$$\vec{r}(t) = (1-t)\langle -1,2,0\rangle + t\langle 3,0,1\rangle$$
$$= \langle 4t-1, 2-2-2t, t\rangle,\ 0 \le t \le 1$$

So, let's get the vector field evaluated along the curve.

$$\vec{F}(\vec{r}(t)) = (4t-1)(t)\vec{i} - (2-2t)(t)\vec{k}$$
$$= \left(4t^2 - t\right)\vec{i} - \left(2t - 2t^2\right)\vec{k}$$

Now we need the derivative of the parameterization.

$$\vec{r}(t) = \langle 4, -2, 1\rangle$$

The dot product is then,

$$\vec{F}(\vec{r}(t)) \bullet \vec{r}'(t) = 4\left(4t^2 - t\right) - \left(2t - 2t^2\right) = 18t^2 - 1t$$

The line integral becomes,

$$\int\limits_C \vec{F} \bullet d\vec{r} = \int_0^1 18t^2 - 6t dt$$

$$= \left(6t^2 - 3t^2\right)\Big|_0^1$$

$$= 3$$

Let's close this section out by doing one of these in general to get a nice relationship between line integrals of vector fields and line integrals with respect to x, y, and z.

Given the vector field $\vec{F}(x,y,z) - P\vec{i} + Q\vec{j} + R\vec{k}$ and the curve C parameterized by

$$\vec{r}(t) = x(t)\vec{i} + y(t)\vec{j} + z(t)\vec{k}, a \le t \le b$$

the line integral is,

$$\int_C \vec{F} \bullet d\vec{r} = \int_a^b \left(P\vec{i} + Q\vec{j} + R\vec{k}\right) \bullet \left(x'\vec{i} + y'\vec{j} + z'\vec{k}\right) dt$$

$$= \int_a^b Px' + Qy' + Rz' \, dt$$

$$= \int_a^b Px' \, dt + \int_a^b Qy' \, dt + \int_a^b Rz' \, dt$$

$$= \int_C P dx + \int_C Q dy + \int_C R dz$$

$$= \int_C P dx + Q dy + R dz$$

So, we see that,

$$\int_C \vec{F} \bullet d\vec{r} = \int_C P dx + Q dy + R \, dz$$

Note that this gives us another method for evaluating line integrals of vector fields. This also allows us to say the following about reversing the direction of the path with line integrals of vector fields.

Fact

$$\int_C \vec{F} \bullet d\vec{r} = -\int_C \vec{F} \bullet d\vec{r}$$

This should make some sense given that we know that this is true for line integrals with respect to x, y, and/or z and that line integrals of vector fields can be defined in terms of line integrals with respect to x, y, and z.

GREEN'S THEOREM

In this section we are going to investigate the relationship between certain kinds of line integrals (on closed paths) and double integrals.

Let's start off with a simple (recall that this means that it doesn't cross itself) closed curve C and let D be the region enclosed by the curve. Here is a sketch of such a curve and region.

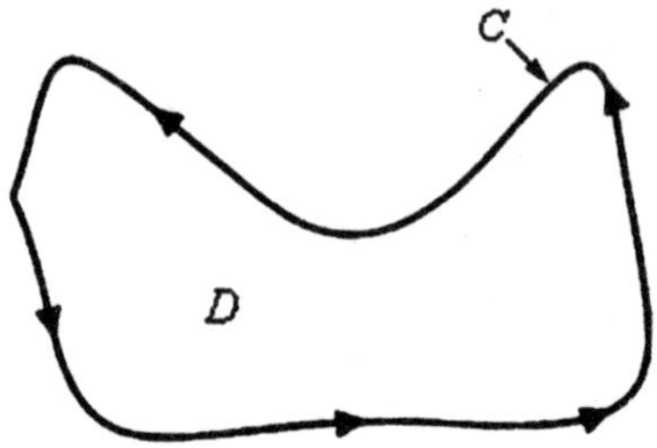

First, notice that because the curve is simple and closed there are no holes in the region D. Also notice that a direction has been put on the curve. We will use the convention here that the curve C has a positive orientation if it is traced out in a counter-clockwise direction. Another way to think of a positive orientation (that will cover much more general curves as well see later) is that as we traverse the path following the positive orientation the region D must always be on the left. Given curves/regions such as this we have the following theorem.

GREEN'S THEOREM

Let C be a positively oriented, piecewise smooth, simple, closed curve and let D be the region enclosed by the curve. If P and Q have continuous first order partial derivatives on D then,

$$\int_C P\,dx + Q\,dy = \iint_D \left(\frac{\partial Q}{\partial x} - \frac{\partial P}{\partial y} \right) dA$$

Before working some examples there are some alternate notations that we need to acknowledge. When working with a line integral in which the path satisfies the condition of Green's Theorem we will often denote the line integral as,

$$\oint_C Pdx + Qdy \quad \text{or} \quad \oint_C Pdx + Qdy$$

Both of these notations do assume that C satisfies the conditions of Green's Theorem so be careful in using them. Also, sometimes the curve C is not thought of as a separate curve but instead as the boundary of some region D and in these cases you may see C denoted as ∂D. Let's work a couple of examples.

Example: Use Green's Theorem to evaluate,

$$\oint_C xy\,dx + x^2y^2 dy$$

where C is the triangle with vertices $(0,0)(1,0),(1,2)$ with positive orientation.

Solution: Let's first sketch C and D for this case to make sure that the conditions of Green's Theorem are met for C and will need the sketch of D to evaluate the double integral.

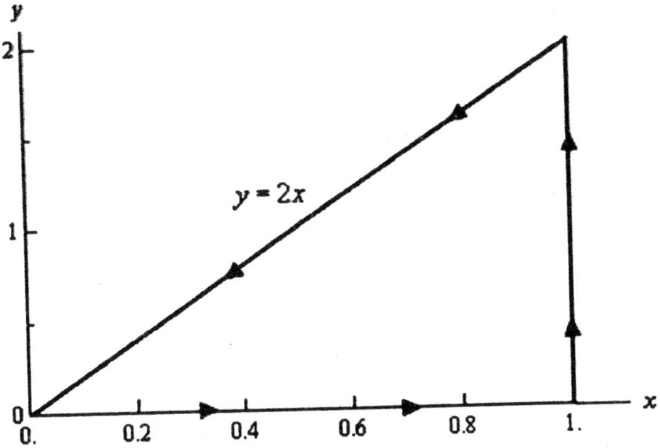

So, the curve does satisfy the conditions of Green's Theorem and we can see that the following inequalities will define the region enclosed.

$$0 \le x \le 1 \quad 0 \le y \le 2x$$

We can identify P and Q from the line integral. Here they are.

$$P = xy$$

$$Q = x^2y^2$$

So, using Green's Theorem the line integral becomes,

$$\oint_C xy\,dx + x^2y^3 dy = \iint_D 2xy^3 - x dA$$

$$= \int_0^1 \int_0^{2\pi} 2xy^3 - x dy\,dx$$

$$= \int_0^1 \left(\frac{1}{2}xy^4 - xy\right)\Bigg|_0^{2x} dx$$

$$= \int_0^1 8x^5 - 2x^2 dx$$

$$= \left(\frac{4}{3}x^6 - \frac{2}{3}x^3\right)\Bigg|_0^1$$

$$= \frac{2}{3}$$

Example: Evaluate $\oint_C y^3\,dx - x^3 dy$ where C is the positively oriented circle of radius 2 centered at the origin.

Solution: A circle will satisfy the conditions of Green's Theorem since it is closed and simple and so there really isn't a reason to sketch it. Let's first identify P and Q from the line integral.

$$P = y^3$$

$$Q = -x^3$$

Be careful with the minus sign on Q!

Now, using Green's theorem on the line integral gives,

$$\oint_C y^3\,dx - x^3y^2 dy = \iint_D -3x^2 - 3y^2\,dA$$

where D is a disk of radius 2 centered at the origin.

Since D is a disk it seems like the best way to do this integral is to use polar coordinates. Here is the evaluation of the integral.

$$\oint_C y^3\,dx - x^3 dy = -3\iint_D \left(x^2+y^2\right)dA$$
$$= -3\int_0^{2\pi}\int_0^2 r^3\,dr\,d\theta$$
$$= -3\int_0^{2\pi} \frac{1}{4}r^4\Big|_0^2\,d\theta$$
$$= -3\int_0^{2\pi} 4d\theta$$
$$= -24\pi$$

So, Green's theorem, as stated, will not work on regions that have holes in them. However, many regions do have holes in them. So, let's see how we can deal with those kinds of regions. Let's start with the following region. Even though this region doesn't have any holes in it the arguments that we're going to go through will be similar to those that we'd need for regions with holes in them, except it will be a little easier to deal with and write down.

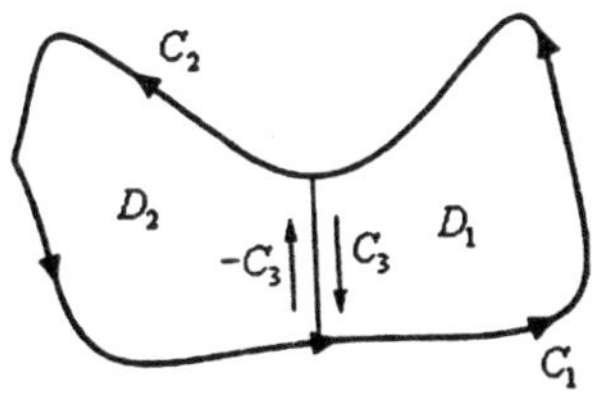

The region D will be $D_1 \cup D_2$ and recall that the symbol is called the union and means that we'll D consists of both D_1 and D_2. The boundary of D_1 is $C_1 \cup C_3$ while the boundary of D_2 is and notice that both of these boundaries are positively oriented. As we traverse each boundary the corresponding region is always on the left. Finally, also note that we can think of the whole boundary, C, as,

$$C = \left(C_1 \cup C_3\right) \cup \left(C_2 \cup \left(-C_3\right)\right) = C_1 \cup C_2$$

ince both $C_3$and $-C_3$will "cancel" each other out.

Now, let's start with the following double integral and use a basic property of double integrals to break it up.

$$\iint_D (Q_x - P_y)\, dA = \iint_{D_1 \cup D_2} (Q_x - P_y)\, dA$$

$$= \iint_{D_1} (Q_x - P_y)\, dA + \iint_{D2} (Q_x - P_y)\, dA$$

Next, use Green's theorem on each of these and again use the fact that we can break up line integrals into separate line integrals for each portion of the boundary.

$$\iint_D (Q_x - P_y)\, dA = \iint_{D_1} (Q_x - P_y)\, dA + \iint_{D_2} (Q_x - P_y)\, dA$$

$$= \oint_{C_1 \cup C_3} P dx + Q dy + \oint_{C_2 \cup (-C_3)} P dx + Q dy$$

$$= \oint_{C_1} P dx + Q dy + \oint_{C_3} P dx + Q dy$$

$$+ \oint_{C_2} P dx + Q dy + \oint_{-C_3} P dx + Q dy$$

Next, we'll use the fact that,

$$\oint_{-C_3} P dx + Q dy = -\oint_{C_3} P dx + Q dy$$

Recall that changing the orientation of a curve with line integrals with respect to x and/or y will simply change the sign on the integral. Using this fact we get,

$$\iint_D (Q_x - P_y)\, dA = \oint_{C_1} P dx + Q dy + \oint_{C_3} P dx + Q dy$$

$$+ \oint_{C_2} P dx + Q dy + \oint_{C_3} P dx + Q dy$$

$$= \oint_{C_1} P dx + Q dy + \oint_{C_2} P dx + Q dy$$

Finally, put the line integrals back together and we get,

$$\begin{aligned}\iint\limits_D \left(Q_x - P_y\right) dA &= \oint\limits_{C_1} P\,dx + Q\,dy + \oint\limits_{C_2} P\,dx + Q\,dy \\ &= \oint\limits_{C_1 \cup C_2} P\,dx + Q\,dy \\ &= \oint\limits_C P\,dx + Q\,dy\end{aligned}$$

So, what did we learn from this? If you think about it this was just a lot of work and all we got out of it was the result from Green's Theorem which we already knew to be true. What this exercise has shown us is that if we break a region up as we did above then the portion of the line integral on the pieces of the curve that are in the middle of the region (each of which are in the opposite direction) will cancel out. This idea will help us in dealing with regions that have holes in them. To see this let's look at a ring.

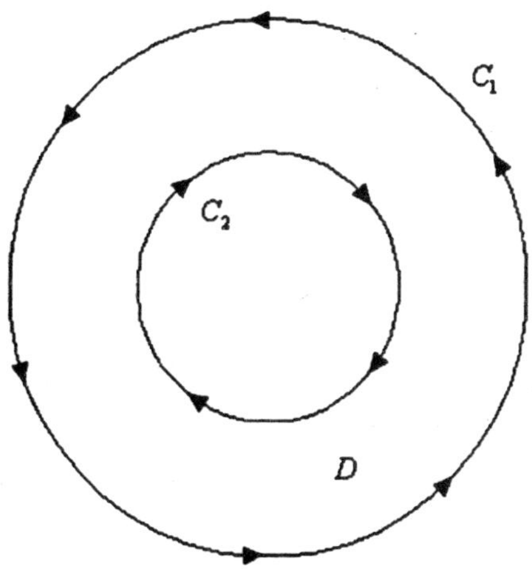

Notice that both of the curves are oriented positively since the region D is on the left side as we traverse the curve in the indicated direction. Note as well that the curve C_2 seems to violate the original definition of positive orientation.

We originally said that a curve had a positive orientation if it was traversed in a counter-clockwise direction. However, this was only for regions that do not have holes. For the boundary of the hole this definition won't work and we need to resort to the second definition that we gave above. Now, since this region has a hole in it we will apparently not be able to use Green's Theorem on any line integral with the curve $C = C_1 \cup C_2$. However, if we cut the disk in half and rename all the various portions of the curves we get the following sketch.

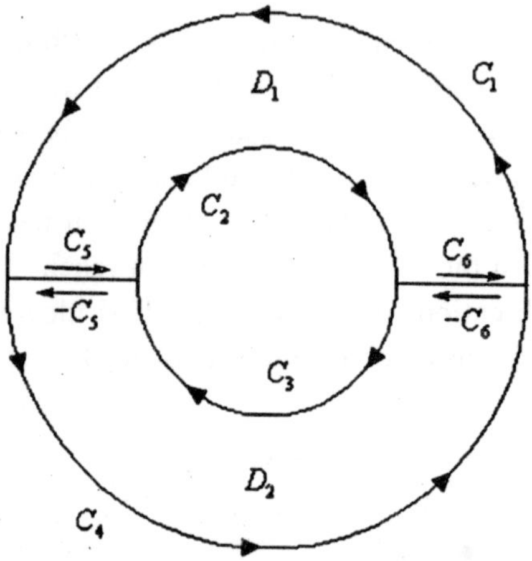

The boundary of the upper portion (D_1) of the disk is $C_1 \cup C_2 \cup C_5 \cup C_6$ and the boundary on the lower portion (D_2) of the disk is $C_3 \cup C_4 \cup (-C_5) \cup (-C_6)$.

Also notice that we can use Green's Theorem on each of these new regions since they don't have any holes in them. This means that we can do the following,

$$\iint_D (Q_x - P_y)\, dA = \iint_{D_1} (Q_x - P_y)\, dA + \iint_{D_2} (Q_x - P_y)\, dA$$

$$= \oint_{C_1 \cup C_2 \cup C_5 \cup C_6} P\,dx + Q\,dy$$

$$+ \oint_{C_3 \cup C_4 \cup (-C_5) \cup (-C_6)} P\,dx + Q\,dy$$

Now, we can break up the line integrals into line integrals on each piece of the boundary. Also recall from the work above that boundaries that have the same curve, but opposite direction will cancel. Doing this gives,

$$\begin{aligned}\iint_D (Q_x - P_y)\,dA &= \iint_{D_1} (Q_x - P_y)\,dA + \iint_{D_2} (Q_x - P_y)\,dA \\ &= \oint_{C_1} P\,dx + Q\,dy + \oint_{C_2} P\,dx + Q\,dy \\ &\quad + \oint_{C_3} P\,dx + Q\,dy + \oint_{C_4} P\,dx + Q\,dy\end{aligned}$$

But at this point we can add the line integrals back up as follows,

$$\begin{aligned}\iint_D (Q_x - P_y)\,dA &= \oint_{C_1 \cup C_2 \cup C_3 \cup C_4} P\,dx + Q\,dy \\ &= \oint_C P\,dx + Q\,dy\end{aligned}$$

The end result of all of this is that we could have just used Green's Theorem on the disk from the start even though there is a hole in it. This will be true in general for regions that have holes in them.

Let's take a look at an example.

Example: Evaluate $\oint_C y^3\,dx - x^3\,dy$ where C are the two circles of radius 2 and radius 1 centered at the origin with positive orientation.

Solution: Notice that this is the same line integral as we looked at in the second example and only the curve has changed. In this case the region D will now be the region between these two circles and that will only change the limits in the double integral so we'll not put in some of the details here. Here is the work for this integral.

$$\oint_C y^3 dx - x^3 dy = -3 \iint_D \left(x^2 + y^2\right) dA$$

$$= -3 \int_0^{2\pi} \int_1^2 r^3 \, dr \, d\theta$$

$$= -3 \int_0^{2\pi} \frac{1}{4} r^4 \Big|_1^2 \, d\theta$$

$$= -3 \int_0^{2\pi} \frac{15}{4} d\theta$$

$$= -\frac{45\pi}{2}$$

We will close out this section with an interesting application of Green's Theorem. Recall that we can determine the area of a region D with the following double integral.

$$A = \iint_D dA$$

Let's think of this double integral as the result of using Green's Theorem. In other words, let's assume that

$$Q_x - P_y = 1$$

And see if we can get some functions P and Q that will satisfy this. There are many functions that will satisfy this. Here are some of the more common functions.

$$\begin{array}{c|c|c} P = 0 & P = -y & P = -\dfrac{y}{2} \\ Q = x & Q = 0 & Q = \dfrac{x}{2} \end{array}$$

Then, if we use Green's Theorem in reverse we see that the area of the region D can also be computed by evaluating any of the following line integrals.

$$A = \oint_C x\,dy = -\oint_C y\,dx = \frac{1}{2}\oint_C x\,dy - y\,dx$$

where C is the boundary of the region D.

Let's take a quick look at an example of this.

Example: Use Green's Theorem to find the area of a disk of radius a.

Solution: We can use either of the integrals above, but the third one is probably the easiest. So,

$$A = \frac{1}{2}\oint_C x\,dy - y\,dx$$

where C is the circle of radius a. So, to do this we'll need a parameterization of C. This is,

$$x = a\cos t$$
$$y = a\sin t$$
$$0 \le t \le 2\pi$$

The area is then,

$$\begin{aligned}A &= \frac{1}{2}\oint_C x\,dy - y\,dx \\ &= \frac{1}{2}\left(\int_0^{2\pi} a\cos t\left(a\cos t\right)dt - \int_0^{2\pi} a\sin t\left(-a\sin t\right)dt\right) \\ &= \frac{1}{2}\int_0^{2\pi} a^2\cos^2 t + a^2\sin^2 t\,dt \\ &= \frac{1}{2}\int_0^{2\pi} a^2\,dt \\ &= \pi a^2\end{aligned}$$

CURL AND DIVERGENCE

In this section we are going to introduce a couple of new concepts, the curl and the divergence of a vector. Let's start with the curl. Given the vector field $\vec{F} = P\vec{i} + Q\vec{j} + R\vec{k}$ the curl is defined to be,

$$\text{curl}\vec{F} = \left(R_y - Q_z\right)\vec{i} + \left(P_z - R_x\right)\vec{j} + \left(Q_x - P_y\right)\vec{k}$$

There is another (potentially) easier definition of the curl of a vector field. We use it we will first need to define the *operator*. This is defined to be,

$$\nabla = \frac{\partial}{\partial x}\vec{i} + \frac{\partial}{\partial y}\vec{j} + \frac{\partial}{\partial z}\vec{k}$$

We use this as if it's a function in the following manner.

$$\nabla f = \frac{\partial f}{\partial x}\vec{i} + \frac{\partial f}{\partial y}\vec{j} + \frac{\partial f}{\partial z}\vec{k}$$

whatever function is listed after the $\nabla$is substituted into the partial derivatives. Note as well that when we look at it in this light we simply get the gradient vector. Using the ∇we can define the curl as the following cross product,

$$\text{curl}\,\vec{F} = \nabla \times \vec{F} = \begin{vmatrix} \vec{i} & \vec{j} & \vec{k} \\ \frac{\partial}{\partial x} & \frac{\partial}{\partial y} & \frac{\partial}{\partial z} \\ P & Q & R \end{vmatrix}$$

We have a couple of nice facts that use the curl of a vector field.

Facts

1. If $f(x, y, z)$ has continuous second order partial derivatives then $\text{curl}(\nabla f) = \vec{0}$. This is easy enough to check by plugging into the definition of the derivative so we'll leave it to you to check.
2. If $\vec{F}$ is a conservative vector field then $\text{curl}\vec{F} = \vec{0}$. This is a direct result of what it means to be a conservative vector field and the previous fact.
3. If $\vec{F}$ is defined on all of $\mathbb{R}^3$ whose components have continuous first order partial derivative and then is a conservative vector field. This is not so easy to verify and so we won't try.

Example: Determine if $\vec{F} = x^2 y\vec{i} + xyz\vec{j} - x^2 y^2 \vec{k}$ is a conservative vector field.

Solution: So all that we need to do is compute the curl and see if we get the zero vector or not.

$$\operatorname{curl}\vec{F} = \begin{vmatrix} \vec{i} & \vec{j} & \vec{k} \\ \dfrac{\partial}{\partial x} & \dfrac{\partial}{\partial y} & \dfrac{\partial}{\partial z} \\ x^2 y & xyz & -x^2 y^2 \end{vmatrix}$$

$$= -2x^2 y\vec{i} + yz\vec{k} - \left(-2xy^2\vec{j}\right) - xy\vec{i} - x^2\vec{k}$$

$$= -\left(2x^2 y + xy\right)\vec{i} + 2xy^2\vec{j} + \left(yz - x^2\right)\vec{k}$$

$$\neq \vec{0}$$

So, the curl isn't the zero vector and so this vector field is not conservative. Next we should talk about a physical interpretation of the curl. Suppose that $\vec{F}$ is the velocity field of a flowing fluid. Then $\operatorname{curl}\vec{F}$ represents the tendency of particles at the point (x, y, z) to rotate about the axis that points in the direction of $\operatorname{curl}\vec{F} = \vec{0}$. If then the fluid is called irrotational. Let's now talk about the second new concept in this section. Given the vector field $\vec{F} = P\vec{i} + Q\vec{j} + R\vec{k}$ the divergence is defined to be,

$$\operatorname{div}\vec{F} = \frac{\partial P}{\partial x} + \frac{\partial Q}{\partial y} + \frac{\partial R}{\partial z}$$

There is also a definition of the divergence in terms of the ∇operator. The divergence can be defined in terms of the following dot product.

$$\operatorname{div}\vec{F} = \nabla \bullet \vec{F}\cdot$$

10

Applications of Integration

INTEGRALS AND AREA

It is an odd fact that the basic ideas of integration predate those of differentiation by nearly 2000 years. Archimedes was quite good at 'doing integrals', albeit in a rather different context. He was interested in the problem of calculating the areas of curved regions.

The ancient Egyptians and Babylonians were quite good at finding the areas of straight-sided regions. Curved regions are an altogether different problem and the ancient Ancients knew little more than that the area of a circle is r^2. More complicated regions need much more complicated ideas. It is not enough just to know the area of a triangle.

Archimedes, in an remarkable anticipation of the methods of the Calculus, tried to adopt a 'limiting values' approach to calculating areas.

His idea was that you could approximate a curved region arbitrarily closely by straight-sided regions and could therefore hope to obtain the area of the curved region as a limiting case of the (calculable) areas of the approximations.

This has since become one of the most basic ways of defining integrals because, as we will see in this section, there is a close connection between the idea of an integral (which we just introduced as the opposite of differentiation) and the idea of an area.

Consider the graph of $y = f(x)$ between $x = a$ and $x = b$. We want to calculate the area between this curve and the x-axis.

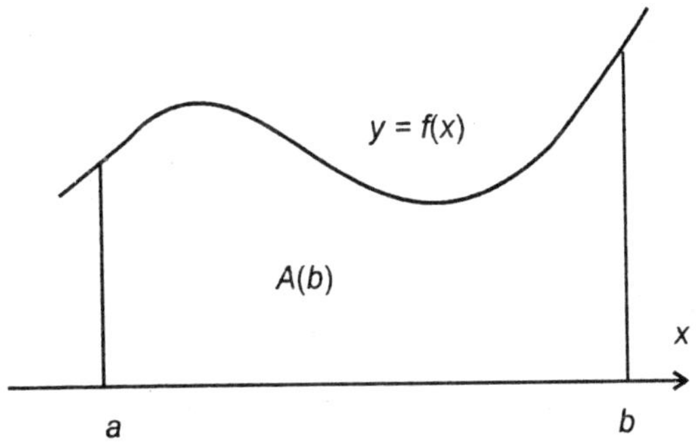

Fig. The Area between a Curve and the x-axis.

In what follows I am going to regard a as fixed and denote the area between a and b by $A(b)$.

Now consider a point slightly to the right of b with coordinate $b + h$.

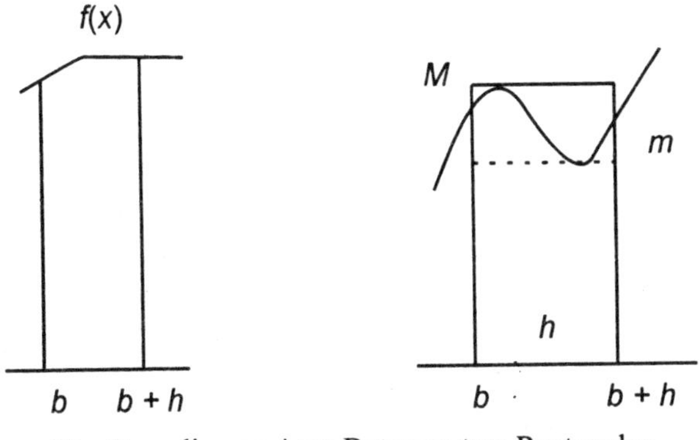

Fig. Bounding an Area Between two Rectangles

Then

$$A(b + h) = \quad A(b) + \text{shaded area}$$

Let me draw a bigger picture of the shaded area. M is the highest point on the graph on this interval and m is the lowest. By looking at the two rectangles in the diagram it is clear that the shaded area has value between hm and hM. So, by above, $mh \le A(b + h) - A(b) \le Mh$ or $m \le \dfrac{A(b+h) - A(b)}{h} \le M$ (we have $h > 0$). Note that the central term is a Newton Quotient.

Now let h shrink down to zero. If f is continuous it is clear that as $h \to 0$ both m and M tend to $f(b)$. The Newton Quotient is sandwiched between them, so

$$f(b) = \lim_{h \to \infty} \frac{A(b+h) - A(b)}{h} = A'(b)$$

So f is the derivative of A. Therefore A is an indefinite integral of f. So

$$\int_a^b f(x)\, dx = \left[A(x)\right]_a^b = A(b) - A(a)$$

but, obviously, $A(a) = 0$. So we get the fundamental result that

$$\int_a^b f(x) dx = A(b)$$

So the area between the graph and the x-axis between $x = a$ and $x = b$ is given by the formula

$$\text{area} = \int_a^b f(x) dx$$

INTERVALS

Before going on to the next section it is worth introducing some standard notation for intervals. An Interval on the real line is all the points between two points (apart from a special case to be done in a moment). These intervals come in a number of types, depending on whether the end-points are included in the range or not.

An Open Interval is one in which the end-points are not included. e.g. $0 < x < 1$. We use the notation (a, b) for the open interval with end-points $a < b$

$$(a, b) = \{x \in R : a < x < b\}$$

A Closed Interval is one in which both end points are included. e.g. $0 \le x \le 1$. We use the notation $[a, b]$ for the open interval with end-points $a \le b$

$$[a, b] = \{x \in R : a \le x \le b\}$$

We also use the notations $(a, b]$ for $a < x \le b$ and $[a, b)$ for $a \le x < b$.

The other types of interval are those that stretch to infinity in one direction or another. For these we use the following notations

$$(-\infty, a) = \{x \in R : x < a\}$$
$$(-\infty, a] = \{x \in R : x \leq a\}$$
$$(a, \infty) = \{x \in R : a < x\}$$
$$[a, \infty) = \{x \in R : a \leq x\}$$
$$(-\infty, \infty) = R$$

Note that things like [-∞, *a*) do not make sense. 'Infinity' is not a number.

The 'area' interpretation of the definite integral gives us the following extra properties:

If $f(x) \leq 0$ on $[a, b]$ then $\int_a^b f(x)dx \geq 0$

(If the graph does not go below the *x*-axis the integral is not going to be negative.)

If $f(x) \leq g(x)$ on $[a, b]$ then $\int_a^b f(x)dx \geq \int_a^b g(x)\,dx$

(If the graph of *f* never goes below the graph of *g* then the area under the *f* graph is greater than that under the *g* graph. That explanation makes sense if both graphs are above the *x*-axis, but the result is true in general as you can see by using the first property on the function $h(x) = f(x) - g(x)$.)

If $m \leq f(x) \leq M$ on $[a, b]$ then

$$m(b-a) \leq \int_a^b f(x)dx \leq M(b-a)$$

(Think of *m* as the constant function $g(x) = m$ and use the earlier results, together with

$$\int_a^b m\,dx = m(b-a).$$

Similarly for *M*.) Finally, a property that is often quite useful

$$\left|\int_a^b f(x)dx\right| \leq \int_a^b |f(x)|dx.$$

GRAPHICAL REPRESENTATION OF THE DEFINITE INTEGRAL

Consider the interval $[a, b]$. Let $f(x)$ be a continuous function defined on this interval. Let us think once more of the problem of finding the area under the graph of $f(x)$ between $x = a$ and $x = b$.

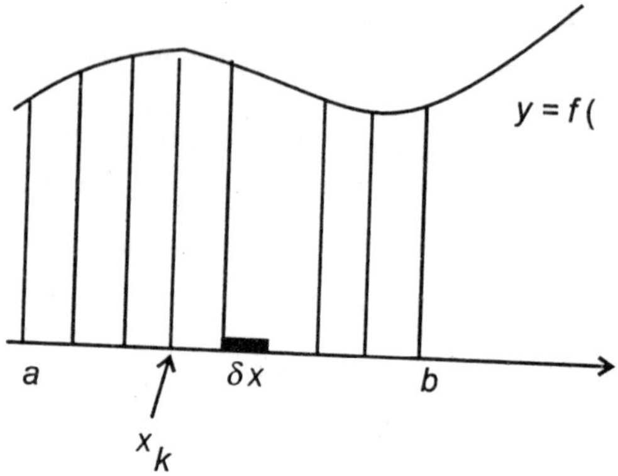

Fig. Dividing up the Area Under a Curve

We will adopt an approach that is much more elementary (and much older) than our previous method. Divide the interval $[a, b]$ up into a large number of small parts. For convenience we will take them all to be of the same width, but that is not very important. Now use this subdivision to break up the area into *thin strips* as shown.

Denote the subdivision points by $x_0, x_1, \ldots, x_n$ where $x_k = a + k\delta x$ and δx is the width of each strip $\delta x = (b - a)/n$.

We can get an approximation to the area under the graph by adding up the areas of the n rectangles shown in the picture. The rectangle on the base $[x_k, x_{k+1}]$ has area $f(x_k)\delta x$. So

$$\text{approximate area} = \sum_{k=0}^{n-1} f(x_k)\delta x$$

Now we use the same kind of argument that we used when

inventing the derivative. As n gets bigger and bigger we expect the sum of the areas of the rectangles to get closer and closer to the true area under the graph.

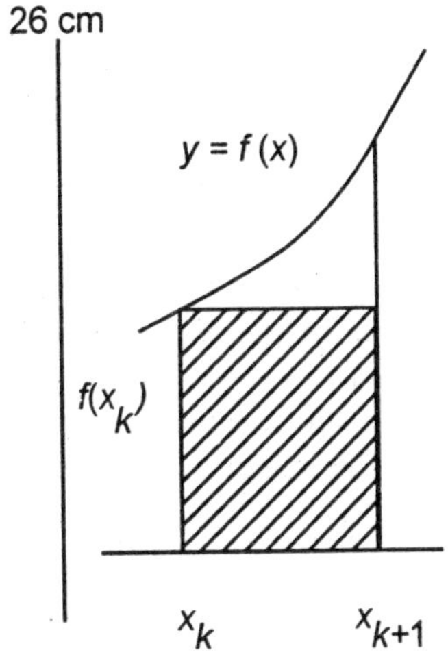

Fig. One of the Rectangles

We would hope that if we took the limit as $n \to \infty$ the sum would tend to the true area as its limiting value. We will assume that this is true. So

$$\text{area under graph} = \lim_{n\to\infty} \sum_{k=0}^{n-1} f(x_k)\delta x$$

But we already know that this area is given by the definite integral. So, putting our two results together, we get (in the case of a continuous function)

$$\int_a^b f(x)dx = \lim_{n\to\infty} \sum_{k=0}^{n-1} f(x_k)\delta x$$

We can in fact take this as a definition of the definite integral (the Riemann Integral), and normally do so in more advanced work. This interpretation of the definite integral is the one that is most useful in applications, as we will soon see.

APPLICATIONS OF THE DEFINITE INTEGRAL

Our new interpretation of the definite integral is very useful. It shows us how to turn a lot of different problems into integration problems. We introduced it in the context of finding areas, but it is vastly more general than that.

Almost anything that comes into the category of 'break it into small bits, approximate on the bits and then add up the bits' yields a definite integral of some form. In this section I will look at a few applications. You may meet others in other subjects.

THE LENGTH OF A CURVE

Suppose we want to calculate the length of the graph of $y = f(x)$ between $x = a$ and $x = b$.

Subdivide $[a, b]$ as before into n small parts. Look at the graph on one of these parts.

The idea is to get an approximation to the length of the graph on this part, in the same way that we used the rectangle approximation when finding the area. Then we add up the approximations and take the limit as n so as to produce an integral which gives the true length.

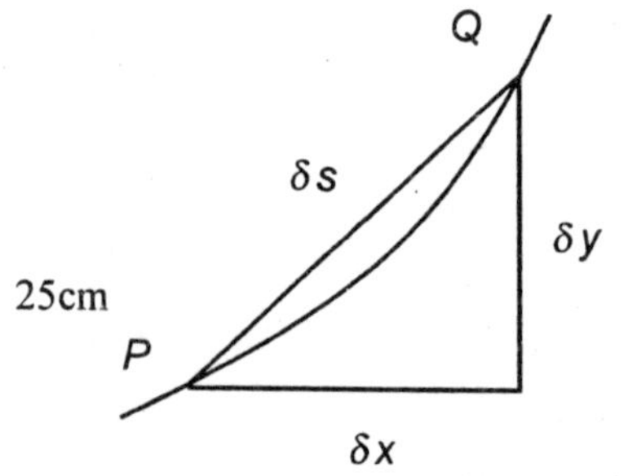

Fig. Straight line Approximation

What approximation do we use?

The obvious approach is to use the length of the chord PQ as an approximation to the length of the graph between P and Q. In the notation of figure this length is

$$\delta s = \sqrt{\delta x^2 + \delta y^2}$$

which we can write more conveniently as

$$\delta s = \sqrt{1 + \left(\frac{\delta y}{\delta x}\right)^2}\ \delta x$$

We now have the approximation to the length which I will write crudely as

$$\sum\sqrt{1+\left(\frac{\delta y}{\delta x}\right)^2}\,\delta x$$

Our new interpretation of the definite integral tells us that, as n this tends to the value of the definite integral,

$$\int_a^b\sqrt{1+\left(\frac{dy}{dx}\right)^2}\,\mathrm{dx}$$

(the $\delta y/\delta x$ is really the Newton Quotient and tends to dy/dx in the limit.)

So we have obtained the formula

$$s = \int_a^b\sqrt{1+\left(\frac{dy}{dx}\right)^2}\,dx$$

for the length of the graph of $y = f(x)$ between $x = a$ and $x = b$.

Note: if the curve is given parametrically by $x = x(t)$ and $y = y(t)$ then a very similar argument gives us the formula

$$s = \int_a^b\sqrt{\dot{x}^2+\dot{y}^2}\,dt$$

for the length of the curve between $t = a$ and $t = b$.

Example: What is the length of the graph $y = \cosh(x)$ between $x = 0$ and $x = a$?

In this case $dy/dx = \sinh(x)$ so

$$\sqrt{1+\left(\frac{dy}{dx}\right)^2} = \sqrt{1+\sinh^2(x)} = \cosh x$$

So the length is

$$s = \int_0^a\cosh x\,dx = [\sinh x]_0^a = \sinh a$$

Example: As a check on whether we are making sense, let's work out the length of a part of a straight line. Let $y = mx + c$

be a straight line. Let us use the formula to find its length between $x = a$ and $x = b$.

$$y' = m \text{ so } \sqrt{1+\left(\frac{dy}{dx}\right)^2} = \sqrt{1+m^2}$$

So

$$s = \int_a^b \sqrt{1+m^2}\,dx = (b-a)\sqrt{1+m^2} = \frac{b-a}{\cos\phi}$$

where ϕ is the angle that the line makes with the x-axis (m = tan). You can easily check that this is indeed the right answer.

Example: What is the circumference of a circle of radius r?

Let the circle be given by the equation $x^2 + y^2 = r^2$. We will get the circumference by working out the length of the top semicircle and doubling the result.

On the top half $y = \sqrt{r^2 - x^2}$. So

$$\frac{dy}{dx} = \frac{-x}{\sqrt{r^2 - x^2}}$$

and

$$\sqrt{1+\left(\frac{dy}{dx}\right)^2} = \frac{r}{\sqrt{r^2 - x^2}}$$

So the length of the semicircle is given by

$$l = \int_{-r}^{r} \frac{r\,dx}{\sqrt{r^2 - x^2}}$$

This is actually a fairly standard integral and you will find it in most tables.

The solution is $dx/ = \int dx/\sqrt{r^2 - x^2}$ arcsin(x/r). You should check this by differentiating the RHS.

$$l = \left[r\arcsin(x/r)\right]_{-r}^{r} = r(\arcsin(1) - \arcsin(-1)) = rp$$

So the total circumference of the circle is twice this, i.e. circumference = $2\pi r$

VOLUMES OF REVOLUTION

Take the graph of $y = f(x)$ on the interval $[a, b]$ and spin it round the x-axis so as to produce what is known as a solid of revolution. We want to get a formula for the volume of this solid.

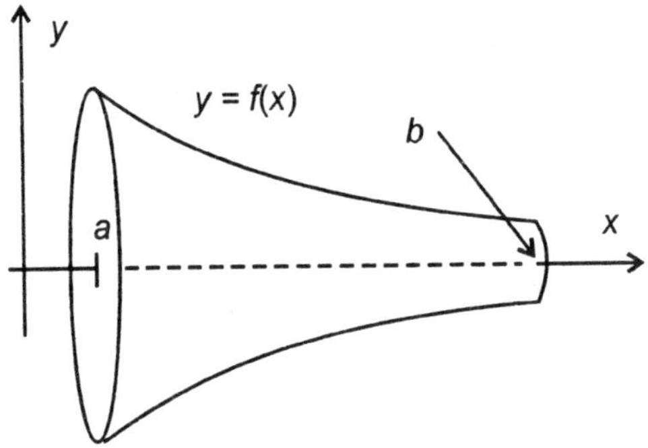

Fig. A Volume of Revolution

The method is almost exactly the same as in the previous examples. Think of the interval $[a, b]$ being subdivided into lots of little bits. Now look at one of the bits and try to get an approximation for the volume of the 'thin slice' of the solid obtained by rotating the piece of the graph on this interval.

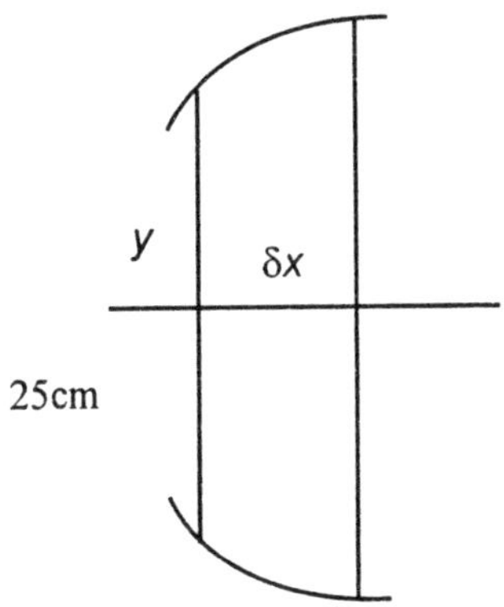

Fig. The Volume of a Slice

In the notation of the diagram, the thin slice of the solid is virtually a *cylinder* of radius y and thickness x. The volume of a cylinder is the product of its height and the area of its base. So we get the approximation

$$\delta V = \pi y^2 \, \delta x$$

for the volume of the slice.

The approximation to the total volume can then be written as

$$V \approx \sum \pi \, y^2 \, \delta x$$

Now take the limit as $n \to \infty$ and get

$$\text{Volume} = \int_a^b \pi y^2 \, dx$$

Example: Cone.

Take the line $y = mx$ on $[0, h]$ and spin it round the x-axis so as to produce a cone of height h and 'semi-angle', α where $\tan\alpha = m$. The volume of this cone is, by our formula,

$$V = \int_0^h \pi m^2 x^2 \, dx = \left[\pi m^2 \frac{1}{3} x^3 \right]_0^h = \frac{1}{3} \pi m^2 h^3$$

If R is the radius of the base of the cone then $m = \tan = R/h$ so, with a bit of rearranging, we get

$$V = \frac{1}{3} \pi \, R^2 h = \frac{1}{3} \text{ base} \times \text{height}$$

Example: Sphere

Take the semicircle $y = \sqrt{r^2 - x^2}$ on $[-r, r]$ and spin it round the x-axis. We get, as our solid of revolution, a Sphere of radius r.

Our formula tells us that the volume of this sphere is

$$V = \int_{-r}^{r} \pi \, (r^2 - x^2) \, dx = \left[\pi r^2 x - \frac{1}{3} \pi x^3 \right]_{-r}^{r} = \frac{4}{3} \pi r^3$$

So the volume of a sphere of radius r is

$$V = \frac{4}{3} \pi \, r^3$$

Example: Consider the funnel formed by taking the curve $y = 1/x$ and rotating it round the x-axis on the interval $[1, a]$, as shown in Fig.

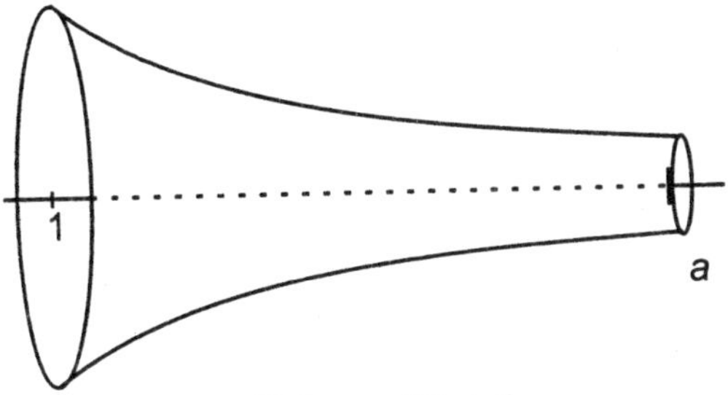

Fig. Volume of Rotation.

The volume of this funnel is

$$V = \int_1^a \pi \frac{1}{x^2}\, dx = \left[-\frac{\pi}{x}\right]_1^a = \pi\left(1 - \frac{1}{a}\right)$$

Now notice that, as $a \to \infty$, this volume tends to the finite value π(because $\pi/a \to 0$).

We write

$$\int_1^\infty \frac{\pi}{x^2}\, dx = \lim_{a\to\infty} \int_1^a \frac{\pi}{x^2}\, dx = \lim_{a\to\infty}\left(\pi - \frac{\pi}{a}\right) = \pi$$

AREA OF SURFACE OF REVOLUTION

I'm going to be brief here and just give you the formula. Suppose we take the graph $y = f(x)$ on the range $[a, b]$ and rotate it around the x-axis as before. Then the *Surface Area* of the surface formed by this is given by area

$$= \int_a^b 2\pi y \sqrt{1 + \left(\frac{dy}{dx}\right)^2}\, dx$$

Example: Sphere

As before, we obtain our sphere by rotating the semicircle

$$y = \sqrt{R^2 - x^2} \quad -R \le x \le R$$

around the x-axis.

For this curve,

$$\left(\frac{dy}{dx}\right) = \frac{-x}{\sqrt{R^2 - x^2}}$$

So

$$\sqrt{1+\left(\frac{dy}{dx}\right)^2} = \sqrt{1+\frac{x^2}{R^2-x^2}} = \frac{R}{\sqrt{R^2-x^2}}$$

So the area is given by

$$\text{Area} = \int_{-R}^{R} 2\pi\sqrt{R^2-x^2}\,\frac{R}{\sqrt{R^2-x^2}}\,dx$$

$$= 2\pi R\int_{-R}^{R} dx = 4\pi R^2$$

So a sphere of radius R has area $4R^2$.

AREA IN POLAR COORDINATES

Read this after you find out about polar coordinates, if you don't know about them already. Suppose we have a curve given in polar coordinate by $r = f(\theta)$. How do we find the area of the region bounded by the curve and the radial lines$\theta = a$ and $\theta = b$?

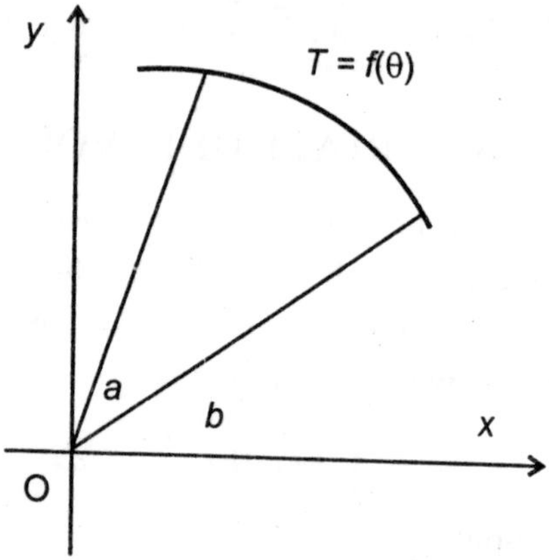

Fig. Describing an area in Polar Co-ordinates

This is another example of the 'thin slices' approach, though a bit different to previous ones. We imagine the θ range from *a* to b being subdivided into lots of small parts of width $\delta\theta$ and the area consequently being divided up into lots of thin pie slices. Any one of these slices, at angle θ, is approximately a triangle with sides $f(\theta)$ and included angle $\delta\theta$.

The area of such a triangle is $\frac{1}{2} f(\theta)^2 \sin(\delta\theta)$. But, for very small angles, $\sin(x) \approx x$. So we approximate the area of the pie-slice by $\frac{1}{2} f(\theta)^2(\delta\theta)$.

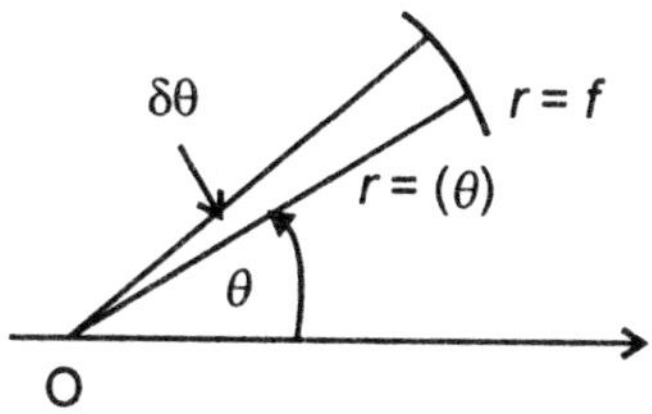

Fig. The Area of a Very Small Segment

Now, 'adding up' these thin slice areas and passing over to the limit we get the formula,

$$\text{area} = \frac{1}{2} \int_a^b f(\theta)^2 d\theta$$

for the area of the region. This is usually written more simply as

$$\text{area} = \frac{1}{2} \int_a^b r^2 d\theta$$

NUMERICAL APPROXIMATION TO DEFINITE INTEGRALS

If a and b are numbers then the value of $\int_a^b f(x)\, dx$ is a number. In many applications we are not interested in the precise value of this number—a good approximation will do. We should be grateful for this because, in practice, very few integrals can be evaluated explicitly so there is no hope of getting an exact answer.

How do we approximate the value without being able to do

the integral? The basic idea is our interpretation of the definite integral as an area (allowing for signs). You could get an approximation to the value of the integral simply by drawing the graph of $f(x)$ on graph paper and counting squares below the graph. This is crude and is a lot of work for a not very accurate result.

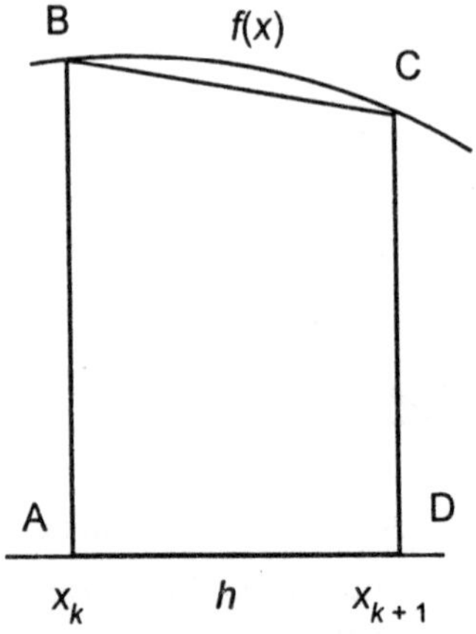

Fig. The area of Below the curve is Approximately the area of the Trapezium

Our thin strips approach to areas gives us a better idea. Choose an integer N and subdivide the interval $[a, b]$ into N equal subdivisions, each of length $h = (b - a)/N$. Call the subdivision points $a = x_0, x_1, x_2, \ldots, x_N = b$. Let the values of $f(x)$ at these points be $f(x_i) = y_i$.

If we look at one strip we can see that a reasonable approximation to its area is given by the area of the 'trapezium' ABCD. The smaller h becomes the better the approximation should be. This area is given by,

$$\frac{h}{2}(y_k + y_{k+1}).$$

Adding up all these pieces we get,

$$\int_a^b f(x)\,dx \approx \frac{h}{2}(y_0 + y_1) + \frac{h}{2}(y_1 + y_2) + \ldots + \frac{h}{2}(y_{N-1} + y_N)$$

This can be tidied up the produce what is known as the Trapezium Rule.

$$\int_a^b f(x)\,dx \approx \frac{h}{2}(y_0 + 2y_1 + 2y_2 + 2y_3 + \cdots + 2y_{N-1} + y_N)$$

The only real virtue of this method is that it is easy the derive (and is rather better than counting squares). Its accuracy turns out to be roughly proportional to $1/N^2$, at least if h is quite small and the graph is not too crazy. This means that doubling the number of strips should about quarter the error.

Let me write down without derivation a much better method known as Simpson's Rule. Same notation as before only this time the number of strips has to be *even*. We will take the number of strips to be $2N$, so $h = (b - a)/(2N)$. Simpson's Rule says that

$$\int_a^b f(x)dx = (y_0 + 4y_1 + 2y_2 + 4y_3 + 2y_4 + \cdots + 2y_{N-2} + 4y_{N-1} + y_N)$$

Note that the pattern of the coefficients is 1, 4, 2, 4, 2, 4, 2,..., 2, 4, 1. This method has accuracy roughly proportional to $1/N^4$, on the same assumptions as before.

In practice these methods are applied by using them repeatedly with increasing values of N until the results seem to have stabilised at the required level of accuracy. Very much a job for computers.

ESTIMATING THE VALUE

This is a footnote. We can use our ideas about integrals and areas to get an idea about the value of the constant e. I quoted a value earlier, but did not deduce it.

Consider the graph of $y = e^x$ between $x = 0$ and $x = a > 0$. We know enough about e^x and its derivatives to be able to say that, on this interval, the graph lies above the tangent line at (0, 1) and below the chord ED.

This says that,

$$\text{area ABCE} < \int_0^a e^x dx < \text{area ABDE}$$

Now ABCE has area $a + a^2/2$ and ABDE has area $a(1 + e^a)/2$. The integral has value $e^a - 1$. So

$$a + \frac{1}{2}a^2 < e^a - 1 < \frac{1}{2}a + \frac{1}{2}ae^a$$

From these two inequalities we get, for $0 < a < 2$, $1 + a + \frac{1}{2}a^2 < e^a < \frac{2+a}{2-a}$

The outer values can be calculated. Put $a = 1$ and get $5/2 < e < 3$. True, but not very precise.

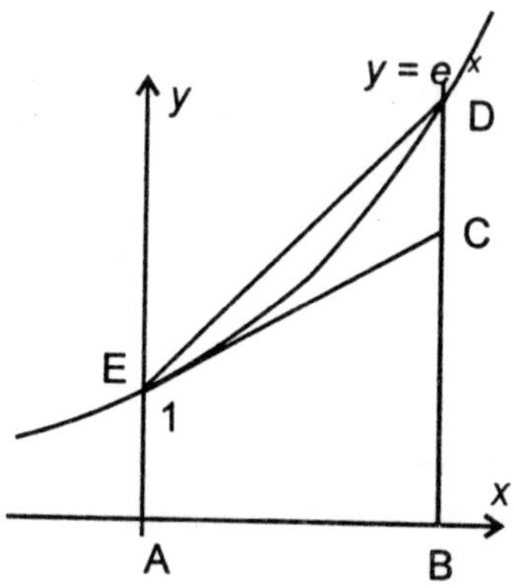

Fig. The Graph of e^x lies between Tangent and Chord.

Now put $a = 1/n$ in the formula: $1 + \frac{1}{n} + \frac{1}{2n^2} < e^{1/n} < \frac{2n+1}{2n-1}$. Finally, raise everything to the nth power: $(1 + \frac{1}{n} + \frac{1}{2n^2})^n < e < \left(\frac{2n+1}{2n-1}\right)^n$.

This is now a good result. The outer expressions can still, in principle, be evaluated by arithmetic. The larger a value we take for n a more precise result we get.

If $n = 10$ we get $2.7141 < e < 2.7206$ (getting close). If $n = 1000$ we get (with the help of a machine)

$$2.71828138 < e < 2.7182821$$

Larger values of n will produce correspondingly better results.

ARC LENGTH

In this section we are going to look at computing the arc length of a function. Because it's easy enough to derive the formulas that we'll use in this section we will derive one of them and leave the other to you to derive. We want to determine the length of the continuous function $y = f(x)$ on the interval $[a,b]$. Initially we'll need to estimate the length of the curve. We'll do this by dividing the interval up into n equal subintervals each of width and we'll denote the point on the curve at each point by Pi. We can then approximate the curve by a series of straight lines connecting the points. Here is a sketch of this situation for $n = 9$.

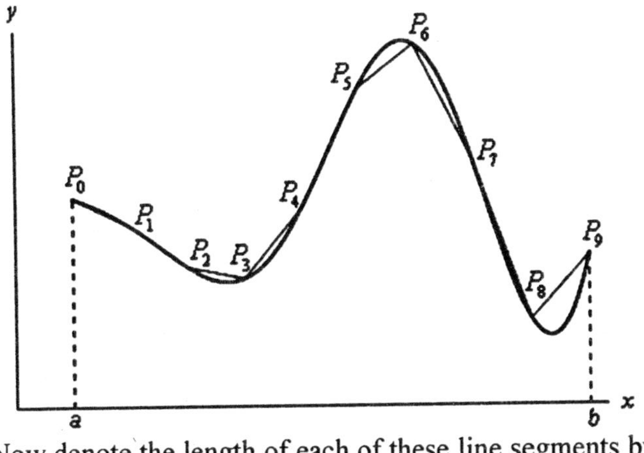

Now denote the length of each of these line segments by $|P_{i-1}P_i|$ and the length of the curve will then be approximately,

$$L \approx \sum_{i=1}^{n} |P_{i-1}P_i|$$

And we can get the exact length by taking n larger and larger. In other words, the exact length will be,

$$L \approx \lim_{n\to\infty} \sum_{i=1}^{n} |P_{i-1}P_i|$$

Now, let's get a better grasp on the length of each of these line segments. First, on each segment let's define Ä$y_i = y_i - y_{i-1} = f(x_i) - f(x_{i-1})$. We can then compute directly the length of the line segments as follows.

$$|P_{i-1}P_i| = \sqrt{(x_i - x_{i-1})^2 + (y_i - y_{i-1})^2} = \sqrt{\Delta x^2 + \Delta y_i^2}$$

By the Mean Value Theorem we know that on the interval $[x_{i-1}, x_i]$ there is a point x_i^* so that,

$$f(x_i) - f(x_{i-1}) = f\ (x_i^*)(x_i - x_{i-1})$$

$$\text{Ä } yi = f\ (x_i^*)\ \text{Ä}x$$

Therefore, the length can now be written as,

$$\begin{aligned} |P_{i-1}P_i| &= \sqrt{(x_i - x_{i-1})^2 + (y_i - y_{i-1})^2} \\ &= \sqrt{\Delta x^2 + [f'(x_i^*)]^2 \Delta x^2} \\ &= \sqrt{1 + [f'(x_i^*)]^2}\,\Delta x \end{aligned}$$

The exact length of the curve is then,

$$\begin{aligned} L &= \lim_{n\to\infty} \sum_{i=1}^{n} |P_{i-1}P_i| \\ &= \lim_{n\to\infty} \sum_{i=1}^{n} \sqrt{1 + [f(x_i^*)]^2}\,\Delta x \end{aligned}$$

However, using the definition of the definite integral, this is nothing more than,

$$L = \int_a^b \sqrt{1 + [f'(x)]^2}\,dx$$

A slightly more convenient notation (in my opinion anyway) is the following.

$$L = \int_a^b \sqrt{1 + \left(\frac{dy}{dx}\right)^2}\,dx$$

In a similar fashion we can also derive a formula for $x = h(y)$ on $[c, d]$.

This formula is,

$$L = \int_c^d \sqrt{1+[h'(y)]^2}\,dy \quad \int_c^d \sqrt{1+\left(\frac{dx}{dy}\right)^2}\,dy$$

Again, the second form is probably a little more convenient. Note the difference in the derivative under the square root! Don't get too confused. With one we differentiate with respect to x and with the other we differentiate with respect to y. One way to keep the two straight is to notice that the differential in the "denominator" of the derivative will match up with the differential in the integral. This is one of the reasons why the second form is a little more convenient.

Before we work any examples we need to make a small change in notation. Instead of having two formulas for the arc length of a function we are going to reduce it, in part, to a single formula. From this point on we are going to use the following formula for the length of the curve.

ARC LENGTH FORMULA

$$L = \int ds$$

Where,

$$ds = \sqrt{1+\left(\frac{dy}{dx}\right)^2}\,dx \quad \text{if } y = f(x),\ a \le x \le b$$

$$ds = \sqrt{1+\left(\frac{dx}{dy}\right)^2}\,dy \quad \text{if } x = h(y)\ c \le y \le d$$

Note that no limits where put on the integral as the limits will depend upon the *ds* that we're using. Using the first *ds* will require *x* limits of integration and using the second *ds* will require *y* limits of integration. Thinking of the arc length formula as a single integral with different ways to define *ds* will be convenient when we run across arc lengths in future sections. Also, this *ds* notation will be a nice notation for the next section as well.

Now that we've derived the arc length formula let's work some examples.

Example: Determine the length of y = In (sec x) between $0 \le x \le \frac{\pi}{4}$.

Solution: In this case we'll need to use the first ds since the function is in the form $y = f(x)$. So, let's get the derivative out of the way.

$$\frac{dy}{dx} = \frac{\sec x \tan x}{\sec x} = \tan x$$

$$\left(\frac{dy}{dx}\right)^2 = \tan^2 x$$

Let's also get the root out of the way since there is often simplification that can be done and there's no reason to do that inside the integral.

$$\sqrt{\left(\frac{dy}{dx}\right)^2} = \sqrt{1+\tan^2 x} = \sqrt{\sec^2 x} = |\sec x| = \sec x$$

Note that we could drop the absolute value bars here since secant is positive in the range given. The arc length is then,

$$L = \int_0^{\frac{\pi}{4}} \sec c\, dx$$

$$= \text{In } |\sec x + \tan x| \Big|_0^{\frac{\pi}{4}}$$

$$= \text{In } \left(\sqrt{2}+1\right)$$

Bibliography

Alexandrovich, V.: *A Treatise on the Integral Calculus: With Applications,* Athens: U of Georgia Press, 2005.

Arthur, J.: *American National Biography: McCutcheon-Moskowitz,* London: Free Press, 1992.

Augustus, D.: *Elementary Illustrations of the Differential and Integral Calculus,* London: Prentice Hall, 2003.

Cajori, F.: *The Teaching and History of Mathematics in the United States,* New York: Addison-Wesley Press, 2003.

Christopher, M.: *American Highway Engineers' Handbook,* London: Manning Publications, 2003.

Frank, M.: *The New International Encyclopaedia,* US: Yourdon Press, 1996.

Friedhelm, E.: *Differential and Integral Calculus,* UK : Addison-Wesley Press, 1995.

Ghosh, J.: *How to Learn Calculus of One Variable,* New York: Addison Wesley Press, 2003.

Hans, N.: *Miscellaneous Notes and Queries,* Urbana: University of Illinois Press, 1998.

Henrie, W.: *Canadian Mathematical Bulletin,* Madison: University Wisconsin Press, 1996.

Jahnke, T.: *A History of Analysis,* New York: John Wiley & Sons Publication, 2001.

Joseph, E.: *A Treatise on the Integral Calculus with Applications,* New York: Addison-Wesley Press, 2003.

Joseph, T.: *The Universal Dictionary of Biography and Mythology*, New York: Yourdon Press, 1981.

Michael, G.: *The First Nonlinear System of Differential and Integral Calculus*, London: Prentice Hall, 2003.

Rao, S.: *Differential Equations*, New York: Addison-Wesley Press, 2003.

Richard, E.: *Introduction to Mathematical Fl id Dynamics*, New York: Oxford University Press, 1989.

Thomas, J.: *Universal Pronouncing Dictionary*, London: Blackwell Scientific Publications, 1994.

Vladimir, B.: *Circular of Information: Issue, Columbia*, New York: Univ. Press, 1991.

Vladimir, S.: *Feynman Integral Calculus*, London: Prentice Hall, 2001.

Walter, Y.: *Encyclopaedia Britannica: A New Survey of Universal Knowledge*, New York: John Wiley & Sons Publication, 2001.

Zhivetin, B.: *Advanced Calculus: Lectures*, KY: University Press of Kentucky, 1996

Index

N

O

P

Q

R

S

T

U

V